Joachim F. Richter

ANTIQUE ENAMELS
for
COLLECTORS

Schiffer Publishing Ltd

1469 Morstein Road, West Chester, Pennsylvania 19380

Contents

Translated from the German by Dr. Edward Force,
Central Connecticut State University.

Copyright © 1990 by Schiffer Publishing.
Library of Congress Catalog Number: 90-61804.

Printed in the United States of America.
ISBN: 0-88740-261-5

This book originally published under the title,
Email Raritäten sammeln,
by Verlag Laterna Magica, D-8000 München 71,
© 1990. ISBN: 3-87467-411-8.

The Fascination of Enamel

Even the pharaohs of ancient Egypt were touched by the magic of enamel, wearing enameled bracelets as well as artistically formed pectorals hanging from colorful jeweled chains on their chests. During the course of its history, enamel has experienced many stages of development and has been cultivated by ever-growing numbers. Though at first only the enameling of smooth surfaces was practiced, great progress was made when rounded and three-dimensional materials and objects could be enameled, which influenced its potential for use and creativity vastly. In the antique world it was Egypt that already knew enamel 1300 years before Christ. Thus a hawk pectoral of Cloisonné enamel from the time of Tutankhamen has come down to us, as has a glass-enamel cup dating from 1470 B.C. Some 500 to 1000 years later, artistic enamelwork was created in Babylon, as it was in Byzantium in the first century A.D.

In the Germanic area, on the other hand, enamelwork of circa A.D. 200 has been found—and referred to by historians by the unflattering name of "barbarian enamel", which should not be taken as a negative connotation to its quality.

Using the pit-enameling technique, the French city of Limoges made a name for itself in the 13th Century, and set new styles in the latter half of the 15th Century with the invention of painter's enamel. Limoges painter's enamel became world-famous, particularly the type known as *Grisaille* enamel.

Shortly afterward, outstanding artists in Blois, the former residence of the French kings, and as of 1670 in western Switzerland created enamel decorations for pocket watches, produced works of breathtaking beauty over a period of several centuries. Miniature figures and landscapes reflected all the finesse of the enameling art. Opaque and translucent enamels, for example, were charmingly combined, precious stones and pearls were added as additional decorative elements to win even higher respect for the enamelwork.

Even though glass enamel from southern Russia had attained an undeniable significance by the turn of the millennium, it was only the metal enamel of Tsarist times that really reached a high point. Fabergé and his masters created the most highly developed works of art. Splendid enameled boxes, vodka cups, chests, snuffboxes, spoons, crosses, picture frames and many other articles appear in museums as evidence of the human creative spirit. Probably the favorite subject of this era was the egg, which has been made with the most varied decor, in many sizes, forms, and enameling techniques.

Just as China can look back over a long tradition of enamel production (beginning in about the 14th Century), present-day enameling in China is enjoying another high point. The world market is supplied with a rich array of jewelry such as necklaces, chains, brooches and rings, as well as everyday utensils from bottle openers and artistic table clocks to enameled ball-point pens and much more. All of this can be bought at remarkably low prices, particularly in specialty shops for Chinese enamel. And when one also knows that, for example, the production of a single Cloisonné egg requires ten manual work processes, as will be shown in this book in reference to a demonstration set, then one can only feel respect and amazement, and wonder whether the price has any relationship to the artistic skill involved. This skill is also expressed in window enamel with its complicated technique, a process that, while it existed previously, has found its master today. Many other techniques and creative elements make present-day enamelwork as interesting as ever.

All these facts are presented in their full extent for the first time in this book. Just as the collection of enamel is an inexhaustible field, this book surpassed all of its planned limitations while being compiled and became more and more inclusive. Today we are all the happier to be able to offer the collector and other interested persons a book of incomparable variety with significant informational value.

Joachim F. Richter
Author

Enameling Techniques

What is enamel?

To be able to appreciate the whole significance of an enameled object and evaluate it as an entity, it is necessary from the beginning to be informed sufficiently about the variety and difficulty of the various enameling techniques and the demands on the artisan's and artist's ability to do enamelwork.

A Special Form of Glass

It is surely known that enamel is a special form of glass, but specific definitions will make the recognition of enamel more authoritative.

To define glass: The main component of glass is quartz, a material found everywhere on earth. But since quartz melts only at the very high temperature of 1700 degrees Celsius, substances that melt more easily, called *fluxes*—particularly sodium, lead and borax salts as well as chalk and small amounts of clay are added to it to make glass. The proportions of these fluxes to the heavy quartz and their specific properties determine the appearance of the glass thus produced, its later chemical and physical properties, and its stability. Thus it is understandable that pure quartz glass is most resistant to corrosion, and so the more fluxes it contains and the lower its melting temperature is, the more sensitive to corrosion it is.

This is the reason why enamel has a relatively sensitive consistency; it not only shatters like glass when struck or subjected to other strong effects, but also is relatively sensitive to corrosion, because it has to melt at about 800 degrees Celsius.

Definition

The literature includes, among others, the following definition of enamel: "Special glasses that are melted in a thin closed layer on a metal surface" (see H. Kyri, "Email", Vol. 10, pp. 435 ff.)—or: "Enamel is a mass of glass that can be melted on metal at about 800 degrees Celsius. Since ordinary glass detaches from the metal surface after the cooling that follows firing, enamel glasses are given additives of substances that increase melting and adhering ability" (see Curt Hasenohr, "Email", p.9).

To determine whether an article was actually enameled or not, first of all the layer applied to the object must really consist of glass, and second, an intimate bonding of glass and metal must have developed during the melting process. This definition is especially important in view of the history of enamel and, for example, in the question of whether or not the technique of enameling was already known in ancient Egypt (see "Geschichte des Email", p.18).

Most artistically enameled articles have copper as the carrier of the enamel. In any case, the famous production sites in Limoges used copper as the carrier when the enamel was opaque and not translucent.

The Limousin Recipe

Fortunately, a recipe from the 16th Century Limousin workshop is known to us. J. A. Graf cited it in the speech he gave on metal as a carrier in 1974 (IIC Congress): "One must have mineral water such as is used there and in which the copper or brass is cooked on which you want to enamel. Be aware above all that neither white nor transparent enamel is suitable, and that those that are applied have to be saturated with wine vinegar. Be sure you do not overheat them or use borax. Clean enamel only in urine or lye. Enameling on brass: The normal way is to engrave deeply first and then cook it three or four times in water with alum. Then dry it well and brush it with a brush such as goldsmiths use. Coat it with enamel and put it in the fire. When it is taken out of the fire, it must be covered immediately with a beaker or something similar so it does not break to pieces."

The second part of this old Limousin recipe refers more to that which we now call deep-cut enameling or *email transparent sur relief."* Thus it was important that the engraved surface shines through transparent (translucent) enamel and allows any engravings to remain visible.

Enamel Carriers

The actual Limoges enamel painting, on the other hand, used copper as the carrier, as described in the first part of the recipe.

From all of this it is obvious how vital the precise mixture of enamel was for the success of any work. The big workshops certainly prepared their own enamels and carefully protected the recipes for their preparation and the careful mixing of the glass masses. The enameler also had to keep in mind that different-colored enamels also had different melting temperatures, for otherwise the different layers of enamel could not have been applied over each other and fired. In addition, enamels of different colors had different specific gravities. Thus some enamels could be applied over a background enamel without sinking into the already enameled underlay when melted. These differences in specific gravities are actually put to use in Grisaille enamel (see p.13) to achieve different gradations of gray to white overlays on the painted or drawn structure.

Like all glass in general, all enamel glass is inherently transparent or, as we shall call it from here on, translucent. Only by adding tin oxide or bone-ash are enamel masses made untransparent or, as we shall call it from here on, opaque. A white glass mass darkened by tin oxide is the basis of all opaque colored enamels. We know enamels in the most varied colors, but glass can be colored only by metal oxides. Translucent and opaque enamels are colored by the following metal oxides:

Yellow: by antimony and iron oxides
Red: by iron oxide, copper compounds and/or cassia purple
Orange: by mixing red and yellow
Green: by copper oxides
Blue: by cobalt compounds
Violet: by manganese oxides
Brown: by iron oxide
Black: by mixing iron and manganese oxides

As noted above, these various mixtures naturally cause small changes in the melting temperature of the corresponding glass and develop the most varied specific gravities in the corresponding glass masses.

Of course we have enamels in the widest assortment of different color shades. Their production methods were the greatest secrets of the individual enamel workshops, which created the most varied color nuances by means of fritting (see p.14).

With the mixtures described above, the enameler produced a wide variety of glass masses. To use them as enamels, the pieces of glass had to be mixed repeatedly by melting and then cooled quickly, to make as many cracks in the glass as possible. These glass lumps then had to be reduced to a powder in a mortar, usually made of agate. This pulverizing was done with the addition of water, which took on a milky coloration. The enamel mass was rinsed in clean water and drained until the water ran clear. Only then had a usable mass of enamel been attained, which could be applied wet or dry.

It must still be added that the French word "Email" (formerly spelled "Emaille"), which is in general use in Germany, is actually of German origin. It arose on a roundabout path from the Latin *smaltum* or *esmaldum* or Italian *smalto*. From this came the Old High German verb *smelzan*, meaning to melt, from which derived the Middle High German term *smalte* or *schmalte*. The French word "Email" probably entered the German language in the 18th Century with the spread of enamelwork from Limoges.

Enameling Techniques

There are four basic types of classic enameling, namely:
1. **Cell enameling,** also known by the French term **émail cloisonné** in Germany.
2. **Pit enameling** or **émail champlevé** or **en taille d'épargne.**
3. **Deep-cut enameling,** also called silver enameling or relief enameling.

Since translucent enamel is used in these enameling techniques, the French term **émail transparent** has also been used, though in the German-speaking area it is more customary to call it *Transparent-Email* (transparent enamel). These three enameling techniques also are called goldsmith's enameling, because they are used mainly in the realm of artistic handicrafts to decorate metal objects, particularly those made of precious metals like gold and silver. On the other hand, there is:

4. **Painter's enamel** or **émail peint,** which is considered a separate art. Painter's enamel might simply be called "Limoges"—because it was invented in Limoges, where enameling in this technique developed into an independent art form.

1. Cell enameling or *émail cloisonné*

The technique of émail cloisonné goes back in its refined form to the Byzantine gold cell enameling technique, which was practiced as sink enameling at the court of Byzantium and developed to a significant point. Theophilus Presbyter described this technique very thoroughly in his "Diversarum artium schedular", in which he says that strips of the thinnest gold are to be laid on the article to be enameled: "Bend and shape these into whatever you want to make in enamelwork, like circles, frames, birds, animals or pictures."

That means that very thin, tall fillets or wires are placed on a copper or gold surface to be enameled, forming the closed cells into which the enamel is placed. In the process, the wires remain as contour lines between the very differently colored enamel cells. This kind of work is very difficult and time-consuming because the wires must consist of a metal whose melting point is far higher than that of the enamel, and must also be soldered to the surface. And it is necessary that the soldering points do not have any holes along the surface, otherwise there would be the danger that various colors of enamel could run into other cells under the wires when they melted, which would very much detract from the beauty of the work.

Cell enameling—which we admire today mainly in Chinese and Japanese work, though enameling as an art was introduced there relatively late (in the 14th Century)—could also be done so that a surface is first covered with a thin layer of enamel, and one on the back as well (*counter-enamel*), and fired. It would then be necessary to let the enameled article cool as slowly as possible, so as to avoid cracks in the smooth enameled surfaces.

The wires are then laid out on the already thinly enameled plate so that they outline the entire picture. Naturally such work can be done only with the steadiest hand and forceps. Then, with just as steady a hand, the already enameled article is put back in the oven, where the background enamel is liquified again. Thus the wires can sink into the liquid enamel. This technique guarantees that there are no holes connecting individual cells, so that no colored enamel can flow from one cell to another during the next firing process.

After the wires are attached without a gap to the surface below and to each other, and after a thorough cleaning, the individual cells can be filled with the desired colored enamel. Depending upon the desired look of the finished work, the cells are either filled to the top of the wires or the wires are left standing a little above the surface of the enamel.

Whether translucent or opaque enamel is used depends on the base materials. Naturally, translucent enamel is always chosen for enameling on a gold background to allow the gold surface, sometimes cross-hatched or chased, to shine through as if by magic—while the color of the chosen enamel gives the shimmering gold background an original, completely new color effect. This kind of enameling arose toward the end of the first millennium A.D., particularly in Byzantium, and probably the best work created was the indescribably beautiful, glowing *Pala d'oro,* a Byzantine cell-enameled work of 1105, which the Crusaders stole from Byzantium, which had later become Islamic, and has been admired since then behind the main altar of San Marco in Venice.

When soldering the wires, one must make sure not only that the wires are set solidly on the base's surface, but also that the "cells" (Cloisons) are sealed, so as to prevent the molten enamel from running into another cell.

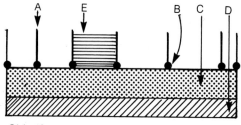

Side view:
A: Wires, B: Solder, C: Base material, D: Counter-enamel, E: Cloison or cell.

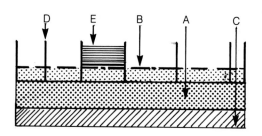

A: Base material, B: Background enamel or fondant, C: Counter-enamel, D: Wires, E: Cell to be filled later (Cloison).
When the wires of the cells are set on the first layer of enamel, which has already been melted once, they will sink into the molten enamel when fired a second time and thus form cells that are completely sealed against the surface as well. The actual cell material is applied only after the second firing.

If the surface of the enamelwork is to be completely plain, then it is ground in the usual way after firing, often after several firings, so that the enamel and the wires are of the same height. The famous Italian sculptor *Benvenuto Cellini* called cell enameling "the true and beautiful art" of enameling.

2. Pit Enameling

The pit enameling, or *émail champlevé*, that we admire particularly in such works as the shrines, votive tablets, and even the crowns of the Middle Ages, is basically a continuation of cell enameling in which the two faces of background material and enameled decor are stressed in a particular way. This makes pit-enameled works much more picturesque. In pit enamelwork, the pits or grooves in which the enamel is to go are cut by engraving, chiseling or etching. This technique allows wider or larger metal bands or bridges to stand between the individual enameled "cells" as individual drawings and deliberate contrasts between the colored enamel surfaces. In pit enameling, the metal surfaces that are to be at the same height as the enamel and form part of the picture are coated with asphalt or resin. The entire rear surface of the metal must be also coated with a material that does not react to acids. Then the metal article, having thus been treated, is laid in an acid bath, which now melts the grooves that are to be filled with enamel out of the metal. The length of the acid bath is determined by how deep the grooves are supposed to be. Then the metal must be cleaned of resin or asphalt and, as described in the Limoges recipe above, be prepared to have the enamel applied. The technique of applying the enamel mass and melting or firing is the same as for émail cloisonné.

Between the middle and the end of the 14th Century, the technique known as deep-cut enameling developed out of émail champlevé.

3. Deep-cut Enameling

Silver plates were usually used for these works. The gold or silversmiths cut drawings with floral patterns, or of people or animals, into the silver plates as sometimes deep, sometimes shallow hatchings. Translucent enamel, usually all of one color, was melted over it in a very thin layer, which filled the drawing cut into the metal and gave it a magical color effect. Its richness depended on the depth of the cuts; the deeper the cut, the darker the color, and the drawing achieved an independent and unique effect. It was not by chance that the technique of deep-cut enameling developed at the time when Gothic glass painting also reached a high point.

3/I. Window Enamel

Also developing from pit enameling, and perhaps also in connection with the rise of glass painting in the 14th Century, was the technique known as *émail à jour* or window enamel. The work was done as follows: The enamelwork made by pit enameling is completely covered with asphalt or resin on the enameled side. Only the back can be laid in an acid bath until the enamel in the pits has been uncovered by the etching acid. If the enamelwork is taken out of the acid bath at just the right time, then the enameled cells are like windows in the remaining metal. These perforations thus resemble cells that have no bottom. In the preparation of window enamel, a metal foil, easily removed after firing, has often been used as a substitute background to hold the enamel. The metal bridges that remain and are connected to each other thus form a framework as well as the contours of the design in *émail à jour*. Naturally, only translucent enamels are used in *émail à jour*, since the effect of such work depends specifically on the light that shines through them. Window enamel has its own particular, enchanting charm because the filtering of the light through the enamel creates unique and fascinating color effects. As noted, this technique was developed in the 14th Century. In the 20th Century it was rediscovered and used in Art Nouveau, where flower petals or butterfly wings, pieces of jewelry and all sorts of bright accessories were made of it.

An example of pit enameling; in this piece, dating to about 1250, the figure and clothing remain raised above the base's surface—only the decorative frame of pits was filled with enamel.

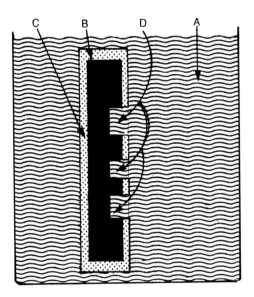

A Base metal for pit enameling in an acid bath:
A: Acid bath, B: Base material, C: Asphalt layer, insensitive to the acid, around the base material; D: The "pits" etched out by the acid.
The acid bath guarantees an even depth of **all** *pits; naturally the pits can also be chiseled or milled out.*

émail à jour

3/II. Email en ronde bosse

The *émail en ronde bosse* must also be mentioned along with the classic enameling techniques. This technique was developed in the 14th Century and has been often used for jewelry, utensils and goldsmiths' figures. The use of émail en ronde bosse required not only artistic talent but also sensitivity for coloring, for this technique allows three-dimensional figures to be covered with enamel. In this process it is especially important to prevent the enamel from flowing off angles and curves and to be sure that, despite repeated firings, the color of the enamel does not change and become dull or dark. Embossed figures that are covered with a very thin layer of enamel in work *en ronde bosse* require the proper application of counter-enamel, a process in which the figure is first enameled from inside to prevent the outer enamel layer from cracking while the article is cooling.

Counter-enamel

A word about *contre-émail* is called for at this point. The back of an enameled plate or the interior of a figure is covered with counter-enamel, usually of low quality, and with color not at all important. Usually counter-enamel is fired at the same time as external or visible enamel. The counter-enamel has the single purpose of keeping the cooling of the metal as even as possible. For example, if the metal should cool more quickly than the melted enamel after the firing process, this would create tension that would cause the newly-melted enamel layer to crack. Since the enamel is heated to a molten condition in the firing oven, the correct use of counter-enamel should be carefully regarded because the melting temperature and duration must be regulated so that the counter-enamel does not run off the surface and thus be unable to fulfill its function.

4. Painter's enamel or *émail peint*

In the history of the enameling art, the French city of Limoges has played an important role on two occasions, once with the pit-enameling process in the 13th and 14th Centuries, and again with enamel painting, which was developed in Limoges in the latter half of the 15th Century. The names of numerous important workshops, to whom we owe the outstanding Limoges work of the 15th and 16th Centuries, have gone down in history:

1. The Monvaërni workshop, late 15th to early 16th Century,
2. The shop of "der Hohen Stirnen", circa 1490-1525,
3. The shop of Nardon Pénicaud, late 15th-early 16th Century,
4. The shop of the Triptychian of Louis XIV, late 15th-early 16th Century,
5. The shop of Jean I. Pénicaud, circa 1510-1545.

In addition, other schools can be named, such as the school of the "violet coats" or that of the "masters of Anäis."

The great progress that these Limoges schools and shops contributed to the art of enameling was the ability to apply layers of different-colored enamels to rather large copper surfaces so that the enamel colors did not run into each other. The enamels were not separated by wires, as is necessary in Cloisonné enameling, but simply had a consistency that kept them over and beside each other despite being melted in the oven, like the paints in an oil painting. Several different processes used in this technique are reported during its scientific history.

One of the processes consisted of first applying a coat of black enamel to the surface, then a coat of white enamel over it. The different application of the white enamel formed the lights and shadows of the design. Often an outline drawing was made on this level, either on the raw white layer, so the black layer showed through, or as a brush drawing on the white enamel.

A different process consisted of covering a copper plate with a solid layer of white enamel and then putting the design on in black enamel with a brush. On this predrawn or prepared enameled surface a variety of colored enamels would be applied with a brush. Translucent enamel was used, when possible, for clothing, decor or backgrounds. Often metal foil, small silver or gold leaves called *paillons*, were also placed under translucent enamel. These were attached to the first layer of enamel with gum tragacanth and then covered with translucent enamel. Such a process was particularly effective if, for example, jewelry, a halo or seams in cloth were to be imitated in an enamel picture. Then the gold or silver shimmered through the translucent enamel and gave the whole effect more of a glitter. Later larger areas were also underlaid with metal foil: In the journal "Restauro", Bettina Landgrebe describes the restoration of a piece of Limoges enamel painting in the Ferdinandeum State Museum in Innsbruck. On the basis of her study, this order of the layers of enamel painting from the first third of the 16th Century was shown. The work is ascribed to the workshop of Jean I. Pénicaud:

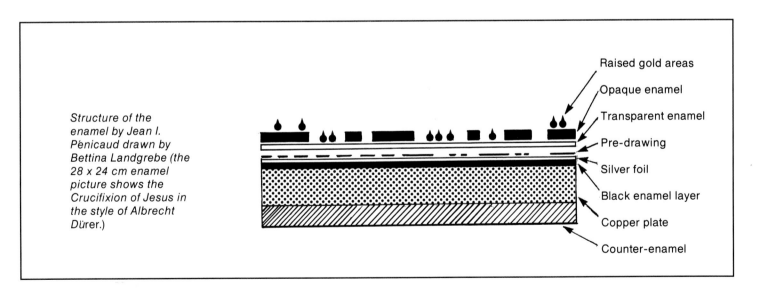

Structure of the enamel by Jean I. Pénicaud drawn by Bettina Landgrebe (the 28 x 24 cm enamel picture shows the Crucifixion of Jesus in the style of Albrecht Dürer.)

Raised gold areas

Opaque enamel

Transparent enamel

Pre-drawing

Silver foil

Black enamel layer

Copper plate

Counter-enamel

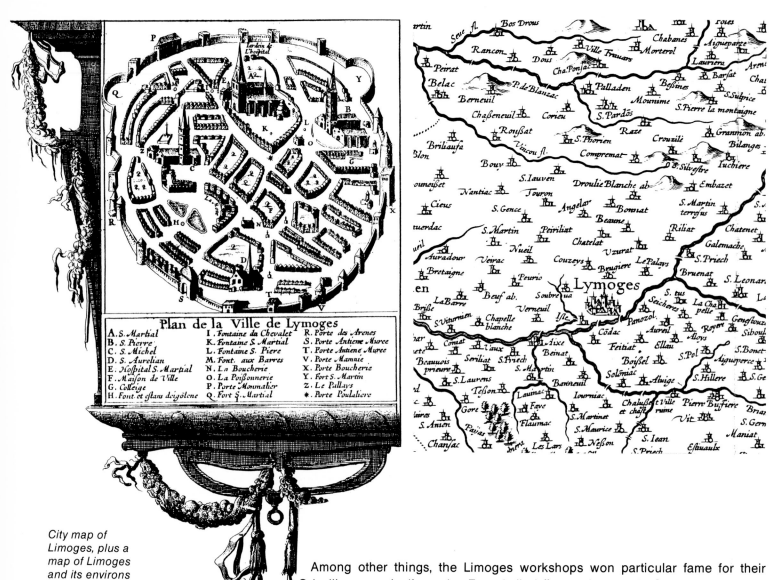

Plan de la Ville de Lymoges

A. S. Martial
B. S. Pierre
C. S. Michel
D. S. Aurelian
E. Hospital S. Martial
F. Maison de Ville
G. Colleige
H. Font. et estans deigolene
I. Fontaine du Chevalet
K. Fontaine S. Martial
L. Fontaine S. Piere
M. Font. aux Barres
N. La Boucherie
O. La Poissonnerie
P. Porte Monmalier
Q. Fort S. Martial
R. Porte des Arenes
S. Porte Antiene Muree
T. Porte Antiene Muree
V. Porte Mamnie
X. Porte Boucherie
Y. Fort S. Martin
Z. Le Pallays
*. Porte Poulaliere

City map of Limoges, plus a map of Limoges and its environs from a "Militaria Gallica" map of 1632.

Grisaille Technique

Among other things, the Limoges workshops won particular fame for their Grisaille enamels (from the French "gris" meaning gray). On a copper plate covered with background and counter-enamel, a relatively thick layer of black enamel is melted. On this black enamel layer the artist drew a picture in white enamel with a brush. The white enamel had a thicker consistency than the black, because of its mixture. When the plate was fired in the oven and both enamels melted, the white enamel easily sank into the black layer. Thus a first white-on-black drawing was made. In a second process, the artist drew the next drawing on the black enamel in white; again the plate was fired in the oven. Now the first white enamel mass sank farther into the black enamel then the second white enamel. If this process was repeated several times, pictures or drawings developed in a variety of gray to white shades. The gray shades naturally differed according to the number of times they had been fired, so that a whole series of different gray shades could come into being. This technique required not only knowledge of the specific gravities, but also exact knowledge of the melting temperatures and times, so as to develop the pictorial effect desired by the artist.

The techniques and talents of the Limoges workshops were so impressive that enamel painting is often called Limoges enamel to this day.

Some of the literature ascribes the invention of enamel painting to a certain Jean Toutin of Chareaudun, circa 1632. This would seem to be incorrect, since enamel painting was done in Limoges well before 1500. Several artists from Limoges traveled to other lands in search of work and spread the technique of enamel painting over all of Europe. A certain Pierre Fromery, who settled in Berlin in the 17th Century, won a good deal of fame. When he signed his works, he specifically noted in his signature that his enamel painting was done on glazed copper plates, as follows: "Fromery à Berlin Kupfer."

Enamel painting in the Limoges style conquered the elegant world of Europe. Copper plates with enamel painting soon decorated clocks, medallions, small cups and boxes, and later complete toilet sets, and complete tea and coffee services. In the 18th Century, for example, the guild of Augsburg silversmiths complained that "enameled goods by the hundreds" were being imported from France and Geneva.

Enamel Painting

Though it does not rank among the classic enameling techniques,

Glass enameling is a separate art form that is older than the literature generally claims.

Firing Temperature

Glass enamels were produced and applied to glass just like all other enamels, with the single exception that glass enamels contain fewer alkalis than metal enamels do. That gives glass enamels a lower spreading coefficient, so that their melting temperature decreases to 700 to 800 degrees C. To paint on glass, the glass enamel powder is rubbed with lavender oil or thick turpentine, so as to stick better when it is painted onto the glass. The opaque colored enamels for glassware are painted on a surface of milky white glass; the translucent colored enamels for glass are applied to transparent lead glass containing boric acid. Matte or flat colors are also used for glass enamel, mixed with tin oxide or porcelain powder—but only when milk glass is being painted to imitate porcelain.

Finally, one more process used in the enameler's art must be mentioned, whose name scarcely anyone knows today, and whose art was surely protected by every enameling workshop as a great secret:

Fritting or Coloring Enamels.

Today we admire the great variety of color shades in many works of enameling art. It has been explained above how the many colors can be produced by adding coloring materials to the basic enamel. It is clear that red is not merely red to us, but exists in many shades of color. But how were these shades created? Scarcely any enameling hobbyist today is able to master the technique of fritting in the home workshop—quite apart from the fact that industry has made it possible to buy a wide variety of colored enamels. But mixed enamel colors are not created simply by mixing two enamels with each other.

Powders of two different-colored enamels mixed with each other produce a surface sprinkled with the two colors after melting, but not a unified new shade. For fritting, the coloring of enamels in the most varied shades of color, a small platinum bowl is needed. One places the pulverized enamel mixture in it to melt it. Since it is impossible to reach into the firing oven with a spoon or stick to stir the melting enamels, once melted the enamel must be cooled quickly so it will break into small lumps of glass. These lumps are then pulverized in the mortar and melted again. This process has to be repeated several times until the two or more colors are blended together so thoroughly that they show only one shade of color.

If, for example, one wanted to turn an ordinary red into what might be called an antique red, one would mix one weight unit of red enamel with ¼ unit of black enamel or black solder.

The many nuances of color that we admire in many enamel works of earlier and also present times came into being this way.

Work with masses of enamel, production of enamel, and techniques have scarcely changed over the course of centuries, even in the 19th Century when the production of enameled objects on an industrial basis became possible. The firing oven simply became larger; larger surfaces of metal were covered with enamel by being dipped in liquid enamel baths or sprayed with powder.

A great secret of every enamel firm was the mixing and grinding of enamel and the correct proportions of the additives. One can read the following text, written by August Bitterling in 1908, with some amusement:

Mixing and Grinding

"The mixing of enamels, that means the combining of all those raw materials whose end product, after melting, is the completed enamel, is done in closed-off quarters. Since the preparation of the enamel is a trade secret, the mixing of enamels in small businesses is done by the proprietor himself or his representative. In larger companies, in view of the great amounts of enamels prepared daily, mixing exclusively on the part of the proprietor or shop foreman is impossible. The following method is suggested:

"The official entrusted with the preparation of enamel personally mixes the smaller components such as sodium, saltpeter, cobalt-nickel oxide, magnesium, brownstone, and also a small part of the borax, feldspar and cryolite called for in the recipe. The worker takes an appropriate quantity of this prepared "pre-mixture" and completes it by adding the right amounts of materials provided for him. A small example of this may suffice.

A white recipe might read:

Borax	....	132 kg
Quarz	....	152 kg
Feldspat	....	130 kg
Soda	....	26 kg
Salpeter	....	6 kg
Kryolith	....	78 kg
Flußspat	....	3 kg
Magnesia	....	6 kg
	Sa.:	533 kg

This can be divided as follows:

			Mixed by the: official worker
Borax	....	12 kg	120 kg
Quarz	....	12 kg	140 kg
Feldspat	....	30 kg	100 kg
Soda	....	10 kg	16 kg
Salpeter	....	6 kg	--- kg
Kryolith	....	28 kg	50 kg
Flußspat	....	3 kg	--- kg
Magnesia	....	6 kg	--- kg
Total		107 kg	426 kg

It goes without saying that completely trustworthy workers are chosen for mixing. As already noted, the raw materials to be mixed must be thoroughly dry and finely mixed. The mixture is thus a more thorough one and the melted enamel is more even in color. For larger-scale work, the use of a portable scale with mixing containers set on it is recommended. The mixing room must be well ventilated. During the mixing itself, the raising of unnecessary dust and the loss of material should be avoided."

Enamel Damage and Restoration

It is proverbial that luck and glass are equally easy to break. Enamel, being glass, is generally exposed to all possible damage that glass can suffer—and being melted onto a foreign object, it also is exposed to all of that base material's usual risks for damage.

The most common cause of damage to enameled objects is external, by being hit, broken or bumped. So many enchanting, pale, and damaged enameled clock faces are proof of that! If the base material changes shape for some reason, the enamel shatters like a broken window. It is a good idea to keep such broken bits of enamel in a small container. If all of the broken-off material is kept, then it may be possible to reassemble it with wax or glue—though not with typical commercial adhesives that would possibly corrode the metal surface. Naturally this requires a great deal of patience, and whether the results are satisfactory or not depends on the talents of the "restorer." The simpler the enamelwork, the more easily slight damage or errors can be repaired, whether by painting over the damaged areas with cold enamel or filling in a crack with porcelain of the same color.

Under no circumstances should one try to repair damage by renewed melting of the whole article (such as in an electric oven), or by new melting of loose enamel pieces followed by reattachment to the base.

Likewise, *under no circumstances* should one try to straighten slight bends or misshapings of the base material!

In both cases the attempted restoration could—or surely would result in the destruction of the whole work, whether out of ignorance of the melting point of the enamel, which could make the enamel darker, spoil its appearance or change its color, or by causing renewed tensions between the base and the enamel, which would make additional pieces of enamel break off! In the world of professional restoration today, the earlier desire and taste for returning slightly damages objects of art to their original condition by "restoration" has been completely rejected.

Slight damage and traces of use, especially in enamelwork, do not ruin either the collectors' value or the beauty of a piece. Cracks in an enamel miniature, breakage in an enameled sign, lack of a cell filling in a Chinese Cloisonné vase will not prevent a collector from treasuring and loving the object—on the contrary, such "wear and tear" show that this particular piece has passed through many loving hands.

Even notable museums keep fragmentary pieces of enamelwork as valuable objects. This is especially true concerning glass enamel—a broken piece with a well-preserved coat of arms from an imperial cup, a broken piece of an apostle cup with a human figure, or the like.

Instead of trying to return it to its original condition, the maintenance of the status quo should be the goal of restorative work. And such work can be entrusted in good conscience only to the professional restorer, who requires not only excellent products for this maintenance, but also knowledge of the chemistry and physics of the object in question. In addition, large museums are accustomed to assist the private collector in any way they can.

Just as annoying, and often just as drastic as damage by external and mechanical influences, is damage that affects the material itself and creates changes.

The penetration of moisture on the edges of the base's material, or in cracks that are a result of mechanical damage, can lead to changes in the base material—to a layer of corrosion on copper, or to rust on enameled utensils made of sheet metal or iron. Such corrosion leads to tension between the metal and the enamel layer.

But different expansion coefficients of enamel layers can lead to tension under the enamel layers too, if the object is exposed to quickly changing climatic conditions.

Mechanical Damage

Caution!. . .

Save it!!

Moisture in Enamel

Is Gluing Allowed?

Caution: Corrosion

In both cases, the resulting tension causes a bubble-like, craterous, upward expansion of one or more layers of enamel.

Enamels, therefore, should be kept dry and not exposed to any abrupt changes in temperature.

Many restorers suggest filling scratches in enamel layers with colored cement to restore the impression of a smooth surface. This is possible with only opaque enamel layers. In a translucent enamel layer over a chased base surface, such cementing would disturb the overall effect rather than improving it.

Any attempted restoration becomes more difficult when enamel layers corrode because of harmful environmental influences, become porous or, on account of its chemical composition, the glass mass reacts in the course of centuries and changes its consistency.

In such cases one must make the very individual decision as to whether the expense of restoration is worthwhile for the preservation of the object or not.

To answer such a question, one should seek the advice of a specialist (preferably from a museum). Many collectors have been surprised to learn that their objects are more valuable than first realized.

The History of Enamel

Did the Ancient Egyptians Know the Technique of Enamelwork?

In any literature on the history of enameling, the question is raised of whether or not the ancient Egyptians of the Pharaohnic era already knew the technique of enamel or enameling. The many decorative objects of the pharaohs, queens and high officials—pectorals, bracelets, fascinating diadems, head rings, decorative collars and even rings with glass stones in silver or gold settings—give the impression that they were made by the enameling technique. Originally the ancient Egyptian goldsmiths and handicraft artists set stone pearls or pieces of jasper, carnelian, turquoise, feldspar or lapis lazuli into the cells of decorative objects. This work was surely very demanding, as the stones had to be worked by chipping or running until they could fit precisely into the cells made for them. They were cemented in with resin that had been mixed with a medium containing chalk. Colored types of cement were often used as well, with the cement being matched to the color of the ornamental stone. In time the high consumption of jewelry in ancient Egypt—just think of how many thousand pieces of jewelry were buried once and for all every year with the embalmed mummies in the burial chambers of the pharaohs, the nobility and high officials—made rare the precious stones needed for jewelry production. And surely the artisans and artists searched for substitutes for these expensive precious stones, seeking a material that could be prepared more or less correctly in form, color, and size for use in the cells.In the New kingdom (as of 559 B.C.) the artisans were able to find a substitute for precious stones by using glass. It could be taken for granted that they got the idea of placing glass in pieces of jewelry in pulverized form and then melting it *in situ;* this would have been possible since the melting point of antique natron-calcium-cilicate glass is lower than that of gold. In addition, lead was found in some types of glass, which could lead to the presumption of a union of glass and metal.

The Egyptian goldsmiths and handicraft workers dealt with prepared molten glass (glass foot or paste) just as they did with precious stones previously. Now, of course, they had the advantage of glass being more or less formable while heated or cooled, which made the work of cutting to fit the cells less difficult. Thus the pieces of glass were fitted to the cloisons after melting and cooling just as the natural precious stones had been, and then glued fast with resin or cement—again colored to match the glass.

Many pieces of jewelry produced in this way have been found in which the cement or resin has aged to the extent of losing its original color and turning to a white lumpy mass.

Didn't the Egyptians know enamel after all, then? Of course! At least one article scientifically corresponds fully to the definition of enamelwork as we know it. This is the *Hawk Pectoral* that was discovered on the mummy of Tutankhamen (1362-1352 B.C.).

Precious Stone Cloisonné

Glass Cloisons

Ancient Egyptian goldsmiths at work; besides the tools, one can recognize the jewelry that is being worked on: collars and pectorals (from a grave painting).

The Hawk Pectoral. . .

. . .is a Genuine Enamel Piece

This ornamental piece, 6.5 cm high and 11 cm wide, is made of gold, lapiz lazuli, and blue, red, and green glass, and portrays the friendly hawk that embodies *Nechbet,* the protective goddess of Upper Egypt. The body of the piece is formed three-dimensionally, while the head was cast separately. The back of the piece consists of smooth gold with chased lines. The wings of the friendly hawk are spread, and the figure holds symbolic objects in its claws.

The significance of this article is that it represents the oldest known use of genuine enamelwork from ancient Egypt.

This ornamental piece has been examined thoroughly, both technoscopically and microscopically, and the fascinatingly brilliant glass inlays have proved to be very much thicker and more precisely set in the individual cells than is the case with most of Tutankhamen's jewelry. In addition, no trace of cement or other adhesive material is found in the cells. Additional proof that this glass was melted *in situ* can be seen clearly in a small pit in the red glass, which was made by an air bubble. This is a further indication that this glass was melted in the gold piece. Bubble holes also can be seen in several places in the blue glass inlays, resulting from their having been melted in the gold. Rough edges also have been found in the gold ribs (bridges), indicating that the entire surface of the piece was smoothed only after firing, after the enamel inlays shrunk slightly into the cloissons during firing. This proves that the *Hawk Pectoral* of Tutankhamen is a genuine piece of cloisonné enamel. It is up to the scientists to examine other pieces of jewelry, at least from the treasures of Tutankhamen, to determine whether or not the Hawk Pectoral is the only genuine enamelwork among Tutankhamen's jewelry. A few of Tutankhamen's rings, at least, look as if they were made also by the genuine enameling technique. No matter how many additional pieces may or may not be found, according to scientific examination, this one piece is sufficient to prove the existence of enamelwork from ancient Egypt, dating to 1362-1352 B.C.

The hawk pectoral of Tutankhamen portrays the Upper Egyptian protective goddess Nechbet; this pectoral is proof that the Egyptian artists, at least at this time, knew and mastered the technique of émail cloisonné.

Glass Enamel

Not only cloisonné enamel can be proved to have been known to the Egyptians; glass enamel too was not only known but used in the times of the pharaohs. One of the first glass vessels with enamel painting (that we know of) is a drinking cup that dates to about 1470 B.C. and is decorated with the sign of Thutmose III (1504-1450 B.C.). The enamel is a yellow type that the Egyptians produced by melting antimony and lead oxide. They also used the same and similar mixtures for the production and coloring of entire glass objects—specifically, yellow and green glasses.

Yellow glass enamels as the Egyptians already knew also appeared on painted tiles in Babylonia, in Mesopotamia, about 1000 years later, circa 500 B.C.

The technique of producing yellow enamel seems to have been forgotten later, for only much later did yellow enamel turn up again, then known as "Neapolitan yellow."

Oddly enough, enamel was practically never used on glass or metal in classic Greece or Imperial Rome. Only in Imperial Byzantium from the 7th to 11th Century, and in southern Russia, did glass enamel appear before and around the first millennium.

Meanwhile Islam had spread through large areas of Europe and the Orient. Islamic artists, forbidden to portray human beings, sought various forms of decorative patterns. So it is not surprising that the Islamic glass artists in Raqqa, Damascus and Aleppo rediscovered the art of glass enameling in the 13th and 14th Centuries and used *luster enamels* to create elegant, decorative hollow glasses of incomparable quality, which are highly desirable among collectors today. Crusaders probably brought much Islamic glass, decorated with enamel, back to Europe as souvenirs. The result was that Franconian and Venetian glassmakers went to Damascus, Aleppo, and Raqqa to study the art of glass enameling there. In 1402 timur declined after the Syrians conquered Damascus; glass painting in Syria suffered a severe defeat. On the other hand, Venice developed a European style of glass painting in the 15th Century, using transparent enamel colors. Such articles also reached Germany, where the Franconian glassmakers produced noteworthy results in the art of glass enameling.

During the Middle Ages, the art of enamel painting on glass declined, at least in Europe north of the Alps, even though the *black solder* had been used there for painting on church windows since the Ninth Century. Black solder, as the name implies, is a black enamel color made from a mixture of colored glass, adhesive, copper, and iron oxide. This black enamel painting was used later on glass and porcelain. In the 15th and 16th Centuries the techniques of enamel painting on glass, and also on Fayence, developed parallel to the techniques of enameling metal. At this time the first glasses with opaque colors on clear glass were painted in Venice. The art of enameling on glass competed with the miniature painting in vogue at the time. At the end of the 16th Century, though, the Venetians began to give up enamel painting on glass, as unpainted cut glass was coming into fashion, depending completely on the "cristallo" for its effect. But north of the Alps, especially in Bohemia, Saxony, Hesse, and Franconia, the art of enamel painting with opaque colors on transparent glass was practiced into the 18th and 19th Centuries. Production of Imperial eagle cups, glasses with coats of arms, hunting, court and apostle glasses attained a true high point, creating valuable collectors' pieces of tremendous value.

A distinctive form of enamel glass painting developed in Nürnberg during the 17th Century, relying on the aforementioned black solder. Glass artists like Johann Schaper painted entire glasses with black solder and then began, after firing, to scratch or rub drawings and shapes out of the black solder.

The 19th Century fondness for decoration gave rise to many mixed forms of cold enamel painting, including on double-walled cups, where the combination of enamel colors and silver or gold foil produced eye-catching effects. While the double-walled cup was preferred in the Biedermeier era, the period of speculation after the Franco-Prussian War brought enamel painting back into style. Many a

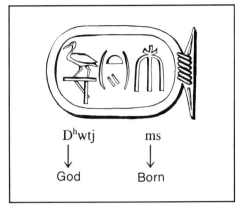

The monogram (cartouche) of Thutmoses, meaning "he who is born of (the god) Toth."

A collector's item with at least a five-figure value: a drinking glass from Syria, late 13th Century, with enamel patterns and inscriptions. Glasses of this type belong to the "Damacene" group.

bourgeois china closet of the late 19th Century preserved and protected enamel-painted wine and liquor glasses, cups and punchbowls; many a neo-Gothic or neo-Baroque vertiko displayed multicolored decorative glasses which were already being produced industrially, then painted in series by capable womens' hands.

Art Nouveau Style

The Art Nouveau Style in Germany, the Secession in Austria, and the Art Deco in France and the USA discovered and adapted molded glass anew in all its sparkling or iridescent nuances, using it to create fascinating shapes with lively enamel decor, intended above all to be new and elegant. Enameled single glasses, vases, and lampshades produced in the large workshops of this era rank among the valuable collections; items often offered by great auction houses today.

In more recent days, workshops have begun to use old recipes and create enamel painting on glass in many striking modern forms.

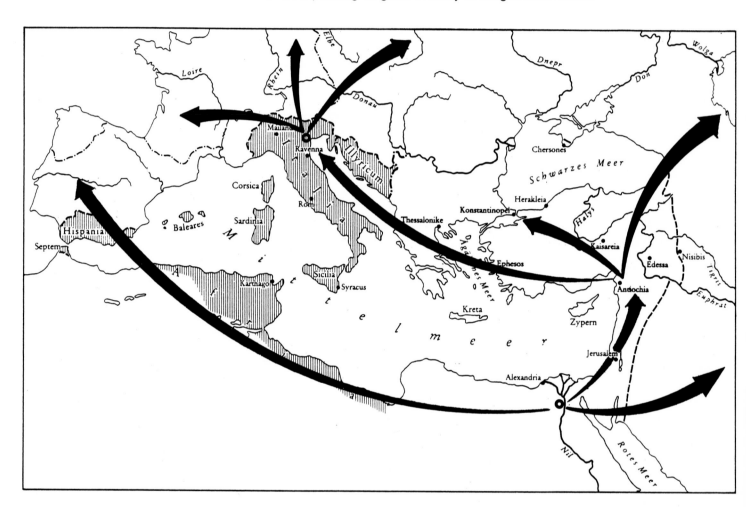

The spread of enamel painting on glass from Egypt until circa 1200.

Enameling in Ancient Greece and Rome

A description of metal enameling in classic Greek and Roman times will be very short, as enamelwork from these times is almost unknown. Enamelwork was still being made in the Egyptian and Nubian area, i.e., the famous bracelet of Queen Amanishahete, dating from the First Century A.D. and found in a pyramid in Meroe'', now owned by the State Museum of Prussian Culture in Berlin. The art of enameling was practiced in Syria; the Goths and Scythians were bringing the art of enameling into the German area even during Imperial Roman times (from which we know it as "barbarian enamel"), but practically no enameled objects of art exist from classic Greece and Imperial Rome. Even though the Greeks painted their stone temples and figures—which is sometimes incomprehensible to us today—no love for metal decorated with enamel existed in these epochs; metal surfaces were preferred that accomplished their effect by means of the material alone. Perhaps we owe to this avoidance of enamel the patina that we admire so much on many pieces of metalwork of the classic Greek and Roman eras.

No Native Enameling Art

To many readers the assertion that the Greek and Roman eras produced almost no enameled works may seem too sweeping a statement. In reality, a few hoards of enameled goods from these times have been discovered. Disputes often arise in the literature as to whether the articles Homer and Hesiod described as being decorated with Electron, such as the ornamented shields described in the Odyssey, were actually decorated with enamel or not.

In ancient Egyptian, "Electron" means an alloy of gold, silver and ???, the formula for which has not yet been rediscovered (and with which the tips of obelisks—as seats of the sun—were decorated). "Electron" also refers to amber in Greek. It is easy to assume that the Electron mentioned by Homer was really made of amber rather than melted enamel. Archaeological discoveries of enameled objects found in ancient Greek and Roman graves were in most cases imported from Nubia, Asia Minor or perhaps also from Cyprus—except for a few pieces of Etruscan origin, which were probably produced in the ancient Italian area as objects of simple, commercially available art. Other simple enameled articles found in the Roman area were made of inexpensive pit enameling on bronze and, in terms of their style, came from the northernmost provinces of the Roman Empire, where enamel—known today as "barbarian enamel", had become common by then.

Electron = Enamel?

From "Barbarian Enamel" to Occidental Gold-Cell Enameling

The term "barbarian enamel" may be appropriate for some simple pieces of jewelry of Germanic or quasi-Germanic iconography. On the other hand, though, as early as 200 A.D. articles with enameled surfaces were made that deserve the admiration they receive today. For example, a candlestick with enamel inlays, dating from the Third Century A.D., that is now preserved in the Cambridge Museum in England, or the colored ornamentation of a horse's bridle, likewise of the Third Century, now in the British Museum.

How strange and amazing the "barbarian enamel" must have seemed to the Romans can be read in the lines of the Second-Century Roman Philostratus: "It is said that the barbarians who live on the ocean cast colors on hot bronze so that they adhere to it, become as hard as stone and retain the patterns formed by them." The knowledge of enameling techniques had probably come to the northern provinces of the Roman Empire from the Sarmantians and Scythians—and in the Seventh Century it experienced another great new influence, coming from the Langobards.

Only thus can it be understood that the German jewelry dating from the time of the migrations is typified by an expressive colorfulness. While the rich and powerful preferred expensive and valuable incrustations of almandine garnets on their jewelry, broador narrow-meshed buckles and clasps for clothing, with blue, green, brown or red cell enameling, were being produced for the less wealthy and the growing class of knights. Garnets and melted glass were even combined on particularly expensive gold-plated clasps.

The most noteworthy new feature of this "barbarian enamel" is the use of pit-enameling processes on base metals such as bronze and copper, with the use of opaque enamels. The ornamentation of these pieces is elementary, primitive, and consists of geometric forms or horseshoe-shaped stripes, rarely of plant designs. Such pieces were usually found in northern Germany, on the Rhine, but also on the Danube, in Hungary, in Denmark and England, and naturally in old Gaul. Along with jewelry such as brooches, pendants, ornaments on weapons and harness, individual containers were also found that showed this type of pit-enameling technique. One of the most famous of these pieces is the sword of Childeric, the father of Chlodwig, King of the Franks, who finally disposed of the last remnants of Roman overlordship in early Gaul. Part of the technical uniqueness of "barbarian enamel" is the fact that on some pieces several colors are united in one field and, though not divided by metal bridges, they are strictly divided from each other. It was even determined from examining several pieces that the artists first placed pieces of glass in the pit cold and then melted colored enamel over them to

An enameled gold disc from Soest. This disc, dating to about 800, was found in a Frankish grave and is in the court museum in Soest today.

attach them. The "barbarian enamel" developed, largely under the influence of Charlemagne, into the great art of occidental gold-cell enameling. Whether the Byzantine cell enameling that developed parallel to it is to be valued more highly is probably just a question of taste, for in terms of evaluating the technical ability the two developments are surely equal. One of the outstanding works of Carolingian goldsmithing art, the Pali otto at San Ambrogio in Milan, shows rich figured enamel medallions with beautifully formed though simply drawn female heads. The bands that link the medallions are alternatingly set with gold enameled plaques and precious stones. One of the most famous enameled pieces of this time is probably the "Iron Crown" in the cathedral at Monza. The ring of this iron crown of the Langobards is set with 24 panels of enameled gold. It is interesting that the crown shows three enameled panels that are probably older—perhaps these pieces were added to the crown during renovation at the beginning of the Ninth Century. With the Carolingian art and the use of cell enameling on shrines, book covers, and reliquary containers, the great age of medieval pit-enamelwork begins. In stylistic terms, both antique and Nordic themes were used, as well as animal designs. This mixture becomes particularly clear on the "Purse of Enger" (presently in Berlin), which one assumes was a pocket reliquary, a gift of Charlemagne to the converted Saxon Duke Widukind on the occasion of his baptism in 785.

In the use of pit enameling and other techniques, it is most interesting that the enamelwork of the Frankish artists shows its own stylistic development, completely separate from that of Rome.

Byzantine Cell Enamel

Rome was lifeless. Ravenna decorated itself with the splendor of its mosaics. The Rome of the East, Byzantium, practiced opulence and luxurious elegance such as can scarcely be imagined today. Just as Napoleon later rediscovered the elegant style of Egyptian art and used it to express the glory and prestige of his imperial crown, the imperial court of the Eastern Empire discovered translucent enamel on a gold background and used its shining splendor to express the glory of rule by divine right. The technique and concept of cell enamel on gold came to Byzantium from the Orient. Still, we know of the earliest cell enameling in Byzantium only from literary sources.

Thus it is assumed that the *gabatina electrina* that Justinian (518-527) gave Pope Hormisdas was an altar lamp decorated with enamel. (The word *electrina* also leaves open the possibility of the use of amber.)

The main altar of Justinian's greatest creation, the Hagia Sophia, was surely decorated with rich enamelwork. In any case, Niketas reported on the destruction of the high altar by crusaders in 1204: "The sacred table, made of various valuable things, which were united by fire and melted to a mass of different colors of the greatest beauty, was smashed to pieces and divided among the warriors."

This undoubtedly refers to costly enamelwork, surely comparable to the later Pala d'oro that is now preserved in Venice.

Naturally Byzantium, or Constantinople, as it came to be called, was not the only location of great workshops that produced Byzantine enamelwork; it can be assumed with good reason—on the basis of stylistic individualities and special work processes—that outstanding workshops and artists of the enameling art were active all over the Byzantine Empire under the influence of the capital city.

Byzantine enamelwork is found with strong Oriental influences, particularly that of Sassanid-Arabic art. In China, in fact, enamel was known as Fo-Lin (a symbolic term used to represent the name "Byzantium"); so the influence of Byzantine enamel art dates back that far. This term makes it clear that China received its knowledge of the "cloisonné technique" from the West.

To be sure, the art of pictorial portrayal suffered a sharp decline in the Eighth Century, including in the Byzantine world. Leo I, the first Isaurian emperor, forbade any honoring by portrait in an edict of 726. For a century, a bloody war was fought

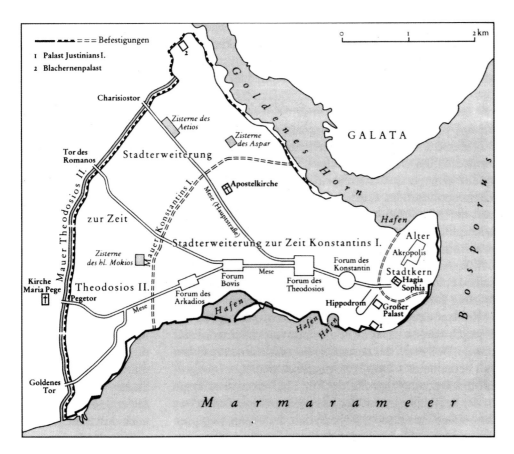

over a religious dogma, which led to the destruction of great numbers of sacred and secular pictures.

In these times many artists not only lost their livelihood but also emigrated to other lands.

Thus it is not surprising that most of the great Byzantine enameled works that have been preserved to the present day were made after the Ninth Century. One of the greatest works made after that time is the already mentioned Pala d'oro, whose creation was surely owed to the art school in Byzantium that was founded by Emperor Constantinus VII Porphyrogenetus (918-956).

The destruction of Byzantium, which had been called Constantinople since 330, by the crusaders caused many outstanding enameled works of art to be brought to Western Europe, and simultaneously ended the great era of Byzantine enamel art. Although the "Byzantines" under Michael VII reconquered their capital in 1261, there was to be no new development of enameling in this city until after 1453, when the workshops in this city must have produced glass enamel again for decorative objects such as lampshades that were used in the new mosques.

From here on, at least, it becomes clear that one independent, systematic history of the enamel art cannot be written. From this point on, the art of enameling has to be considered in the light of general intellectual and artistic history. A treatment of enamel art with no gaps would thus be possible only within the vast concept of a general, thorough history of art and culture.

For this reason the greatest high points of the enameling art will be discussed in separate chapters, though in chronological order.

Thus at this point we must make an excursion into another realm and go to Korea and Japan, where the art of enameling found unique expression.

Justinian, 482(?)-565—seen here in a mosaic portrait—had the Hagia Sophia built.

21

Constantinopel

Constantinople—as it was called until conquered by the Turks in 1453—in one of the oldest pictures of it; from the Schedel World Chronicle of 1493. The Hagia Sophia is directly behind the city wall.

Part of the Pala d'oro, which presently decorates the high altar of San Marco in Venice. This section shows Daniel and Solomon. The figures, formed by the finest cell enameling, stand in front of smooth gold panels that symbolize the heavenly paradise.
The columns between the figures and the garlands in the fields are also made of cell enamel. Additional ornamentation with precious stones decorates the exquisite panels.

Asiatic Metal Enamel: China, Korea and Japan

To begin with, in the Asiatic area Cloisonne and pit-enameling techniques were used equally, side by side, often even together on the same piece. The Chinese enamel artists saw as good as no difference between these two techniques, and many pieces that we could regard superficially as Cloisonné work were really made by the pit-enameling or Champlevé technique. In the process, the pits were cut into the metal, or etched out of it, so close to each other that the fillets between them are so thin that they look as if they are made of soldered-on wire.

These techniques were also enhanced by a typically Asiatic and visually uncommon, elegant, and new technique of enameling, the émail repoussé. In the process the cell walls were formed from the back, embossed into the metal plate, so as to stand out from the front like reliefs. Enamelwork of the Ming era may show all of these techniques used on one and the same object, while in the reign of Emperor Ch'ien Lung the essential element of enameling technique was simply the émail repoussé.

It is remarkable that the Chinese took so long to utilize the techniques of enameling in their art work at all. Not only did they already possess the technical knowledge and the materials as prerequisites for the invention of enameling about the Fifth Century B.C., but by the Seventh Century they must have imported enameled works of art, for the earliest Western pit enameling that has been found in archaeological strata in China comes from this time. These Scythian, or even pre-Scythian, belt buckles could even date from the Second or Fourth Century and certainly reached China via the Caucasus.

Chinese Cloisonné became customary, and in perfected form, at the end of the last Yuan emperor's reign.

It must be noted that scientists dispute this point heatedly, for Japanese experts date two objects, one of them Shoso-in-style mirror with a diameter of 18.2 cm, the other a lotus base for a Buddha figure, as coming from the Tang era, thus at least from the Eighth Century.

Be that as it may, the Chinese term for Cloisonné is Fo-Lin, or Fa-Lang—both are formed from the Chinese name for Byzantium. In other sources enamel is also called *ta shih-yao,* which means "Arabian wares", while *yao* means "burned wares" or oven-fired goods. In other Chinese sources, enameled objects are also known as "burned wares from the devil's land" *(kuei kuo yao).*

Scientists explain these different expressions for the same articles by saying that enamel came to China by two routes, from Byzantium via the land route and from Arabia via the sea route to southern China.

This explanation will not surprise the historian. Toward the end of the 13th Century, the grandsons of Ghengis Khan (or Khakhan) expanded their empire to include all of China and stretch to the borders of the Byzantine Empire, and also occupied Persia, where they, the Mongols, ruled the Illkheans. With that a definitely "bipolar" contact with the world of enamel art was formed.

If one considers the character of Genghis Khan's grandsons, one can surely ascribe to the "Chinesified" and art-loving Kublai Khan the promotion of the new art of enameling, as he was the only Mongol Khan to reside in Peking and was very much interested in all the arts there.

With these facts in mind, it is not surprising that enamel is first mentioned in Chinese literature at the beginning of the 14th Century. The Chinese became conscious of history quite early. In 1387 a certain Tsao Ch'ao wrote the book *Ko ku yao lun* (Discussion of the Essential Criteria for Antiquities). A short chapter in it discusses the "Arabian wares": "The body of the piece, consisting of copper, is decorated with colored patterns that are melted together out of various materials. It resembles the *Fo Lang* inlay work. We have seen incense vessels, flower vases, round boxes with lids, wine cups and the like, but all are intended only for use in the ladies' chambers, for they are too colorful and ornamental for scholars' libraries of restrained taste. They are also called "wares from the devil's land." Today a number of people from the province of Yunan have opened factories in the capital, where

Enamel Repoussé

A cup in the form of the sacral vessel Ku; rich Cloisonné work with lotus tendrils from the Ming era, early 15th Century.

wine cups known as inlay work from the devil's land are made." Thus it is obvious that the Chinese not only learned fast but also could progress somewhat from the "baubles" for the women's rooms; for a Fifteenth-Cntury source—only a century later—already states: "The enamels of similar types that are made for the Emperor's palace are beautiful, glowing and wonderfully made."

It is not known where the first Chinese Cloisonné work originated, but as early as the 17th Century, Emperor K'ang Hsi was already having only enameled works of art produced in one of his 27 palace factories.

It is clear that the knowledge of how to produce beautiful enameled objects spread from China to Korea no later than the 16th Century, and from Korea on to Japan.

The Japanese, according to the literature, used enamel jewelry only for minor purposes until the 19th Century. Some experts dispute whether the aforementioned Shoso-in Mirror was not actually a Korean work, or even an Arabic work of the Eighth Century that reached Japan via China. Goepper describes the mirror as a 17th-Century work.

The Chinese art of enamel reached its first high point in the Ming dynasty. Enamel was especially popular among the Mongols, who preferred loud and glaring colors, and enameled and gilded religious pictures soon decorated their temples. The enameled objects of the Ming era are typified by particularly impressive colors:

A bull with pit-enamel decor from the Ch'ien Lung era (1736-1796); the gold and red enamels highlight the shape of the body.

turquoise and lapis lazuli blue, a dark coral red tending to brown, bright yellow, deep green, black, and white. Such colors on pots, plates, ritual vessels, cups with floral decoration or abstract animal symbols always indicate early works made before the end of the 16th Century.

The *famille rose* colors do not appear until the end of Emperor K'ang Hsi's reign. It is also noteworthy that, while brass wires were used for Cloisonné work in the Ming era, Cloisonné made from the 18th Century on usually used gold wire, and occasionally copper wire.

Famille-rose

Enamelwork of the highest quality was produced in China until the 19th Century. Most of the Chinese Cloisonné pieces to be seen in European or American museums today were stolen from China by Europeans in the latter half of the 19th Century. This happened first in the plundering of Yuan Ming Yuan in 1860, and then to a much greater extent around 1900, when the Europeans believed that the Boxer Rebellion in China had to be stopped by rigorous means.

The Europeans developed a taste for the delicately beautiful art of Chinese enamel, and European workshops competed to copy such pieces in ways that would not have been achieved naturally.

When imperial patronage—which enameling had always enjoyed in China—ended, Chinese enamel also lost its splendid appearance and artistic power to overwhelm. The Cloisonné of the Ming era, thus of the 15th Century, were probably the finest of all enamelwork; Asia never again produced enamelwork of this quality.

A similar development to that in China also took place in other Asiatic lands, particularly in Japan. Naturally various typically Japanese pictorial themes found their way into the illustrative iconography: floral patterns of enchanting grace, emblems and symbols of powerful significance, color shadings as on costly glazing, linear ornamentation with sharp outline drawings.

Of course bewitching Japanese enamels—particularly from the Edo era of the 19th Century—are known to us. Not only were vases and cups decorated with Cloisonné plaques, but Japanese artists, with a sure sensitivity for style and plastic effect, "dressed" already cast or chased figures with complete preformed pieces of clothing; thus arose figures of Samurais or Shinto deities in splendid colorful clothing.

Great works of enamel art were very rare in Japan, for enamel did not play a very large role in Japanese art. The most refined, never again equaled, glazes on clay articles, gold and silver dippings on metal, and perfected lacquer painting on any type of wood were always much more important, preferred, and more typical means of artistic expression in Japan—and sources of more sensitive joy in depiction than the "art from the devil's land" that arrived from China and Korea.

Thus it is not surprising that Roger Goepper, a well-known expert on East Asia, cannot find anything particularly good to say about Japanese enamel art. He reported that at the end of the 18th Century, the Hirata clan had already produced sword-hilts *(tsubas)* with enamel decor that actually look more like industrial products. In a similarly negative tone he described the vessels with Cloisonné decor that were made in the house of Kagi Tsunekichi beginning in 1839. The decorations of these vases decline to the level of lifeless imitations, and the badly applied enamel on most of these vessels is uneven and easily broken. Toward the end of the 19th Century the enamelwork of the Shippo firm in Kyoto was already being produced industrially. Its effect on us today is part kitsch, since glittering gold sprinkles were melted into these enamels. Still in all, Sippo had developed its own technique, though it can be taken lightly. The enamels were applied in the manner of Cloisonné technique—before firing, at least, the only loosely applied wires were removed so the colored surfaces abutted directly on each other. This type of enamel is also called *Musen-Shippo*. All the same, we are indebted to this firm for several creations in émail à jour, in which the metal surface is absent, making the enamel as transparent as glass. The Japanese call this type of enamel *Shotai-Shippo*.

Meiping Cloisonné vase (plum vase) with black background, from the mid-17th Century. Such vases were generally used only for plum blossoms, hence the "plum vase" name.

Cantonese Painter's Enamel

The art of enamel painting on porcelain developed primarily in Canton and came to China from foreign lands; thus it was called Yang Tz'u (porcelain from abroad). Whether this art became known in China before or just after 1700 is not yet known. But a letter from a certain Matteo Ripa, dated 1716 reports the following from Ch'ang Ch'un Yuan:

"Since His Majesty is enchanted by our European painter's enamel, he is trying in every way possible to introduce the process in the imperial workshops which he has set up for that purpose in his palace.

"The result was that something really was accomplished with the customary colors of porcelain painting, and with the help of several large enameled works which he obtained from Europe. But since he also wanted to have European painters, he commanded me and Castiglione to paint in enamel."

But Matteo Ripi and Castiglione refused to paint indefinitely like slaves in the emperor's workshop.

What becomes clear from this episode is not just how desirable enamel painting on porcelain was in China, but that European stylistic influences had been accepted with the popularity of porcelain enamel painting. This enamel painting must have attained considerable success at the right time in China too, for as early as 1721 the Chinese Emperor gave the Russian Tsar a service of golden bowls with enamel painting; the same emperor, Ch'ang Hsi, sent an even larger group of gifts to Pope Clement I in the same year, including ten vases with enamel painting.

These facts, attested to by the repeated contact between the Vatican and Peking, show that enamel painting on porcelain probably became known in China through Christian missionaries, chiefly Jesuits. One of the most famous enamel painters of the Chinese-European style was Father Rinaldo Maria di San Giuseppe, who held

A flat bowl with flowering twigs, from the Tung Cheng era; a lovely example of Cantonese painter's enamel.

the position of overseer of handicrafts and art work at the court of the Thirteenth Prince. The prince wanted to get more European artisans than he already had, and Father Rinaldo suggested to him that he summon a trained *smaltista,* (enamel painter) from Europe. This Thirteenth Prince later became the Emperor Ch'ung Lung, under whom enamel painting in *famille rose* style was developed to its greatest height. The *famille rose* plates of thin eggshell porcelain with ruby red undersides rank today among the most desirable of all collectors' items. This porcelain was, in fact, produced elsewhere but painted with enamel colors in Canton. Canton enamels often bear the mark of the emperor's reign on their underside. They stand out above all in being painted on a background of usually opaque enamel. Thus they are similar to the Delft Fayence, with its painting on tin glazes. In Canton enamel the enamel colors are applied to the unfired soft porcelain and sinks into the soft surface when fired, thus achieving a softer-looking effect.

Enamel colors that were applied to the porcelain glaze, on the other hand, stand out as relieflike incrustations.

Of the Cantonese shops, the best-known is the Des or Pai Shih. This signature is often found on collectors' pieces along with lines of poetry or short prose lines. It has been asserted, though, that Pai Shih never had his own enameling shop, but was the name of a painter whose pictures were often copied on enamelwork. There are also finely painted plates with flower patterns that bear this signature as well as borders in *seven-stripe* style.

Canton enamels often show a wealth of non-Chinese decorative patterns, an indication that they were often made for export, thus for foreign markets. Enamel-painted vessels were even made in the form of English silver teapots, because they were produced exclusively for England. Even today, low-priced Canton enamel is produced in Canton for export.

As early as the end of Emperor Ch'ien Lung's reign, the quality of Canton enamel had declined greatly, even though a gentle pale blue on a silver background attempted to give these pieces a certain elegance. Present-day Canton enamel, on the other hand, seems to lack any charm or grace.

Eggshell Porcelain

Scarcely believable: this European scene was painted in Canton in the 18th Century. Fine enamel painting like this was made for the European market and was made to suit the tastes of customers there.

Pit enamel from the circle of Nikolaus of Verdun, circa 1200; a standing bishop as the founder of a church.

Pit Enameling of the Middle Ages

In the Thirties, French and German patriots were debating whether the pit enameling of the Middle Ages was a German or French invention. Well, it was neither! The great, impelling developments of medieval pit enamel—definitely a continuation of the refined "barbarian enamel"—took place in Flanders and Brabant, a European area now part of Belgium, to which the art history of Europe and the world owe so many great works. Names that are associated with the beginnings of medieval pit enameling were Reinier von Huy and Godefroi de Claire. Reinier von Huy created the Liege baptismal font and the equally renowned censer of Lille. Notable among the surviving works of Godefroi de Claire are the works that Abbot Wibald of Stablo commissioned from this artist, such as the Alexander Reliquary of 1150, a box in the form of an altar, onto which the chased silver head of the saint was attached and which today is preserved in the Musée Cinquantenaire in Brussels. Only a few half-figures of angels from the "Renaclus Altar", which Godefroi was likewise commissioned to make by the same abbot, and which must have been greatly impressive, now remain in the Sigmaringen museum. Godefroi also worked for churches in Deutz, and the fact that he received a commission from the Church of St. Pantaleon in Cologne indicates that he was a great master of outstanding drawing and sculpting.

Pit enameling has many advantages over cell enameling, particularly the possibility of greater freedom in designing and including large areas of metal in a design. In addition, only opaque colors were used in pit enameling, and they could be produced at that time in much greater quantities than translucent enamel. Then too, opaque colors were not as liquid when fired at that time, so that several colors could be placed side by side in a pit cell without running into each other. That allowed attractive colorings and shadings. Even though smaller works of art were generally produced by pit enameling, it was possible, thanks to the lower cost of the metal, to produce copper articles of great size. Thus the size of pit-enameled articles was almost unlimited, in contrast to cell-enameled objects. Consider above all the sacred works of the early Middle Ages, as well as the Hohenstaufen era; these works express the entire nobility and heavenly detachment in the minds of the pious in those times, an expression of the contemporary theological teaching.

Between the first milennium and the 14th Century, the name of the city of Limoges stands out. One of the most enchanting, iconographically impressive, exquisitely formed, and best-proportioned art works of the Limoges School is surely the "Casket of the True Cross," presently preserved in the basilica at Toulouse. The most interesting feature of this work is the fact that the figures portrayed on it are not formed in the enamel but with gilded chased copper plaques, while the heavenly, transcendental background glows with a gentle silky luster of opaque enamel colors and great variations of color.

It might be noted that only the heads of the figures were made as three-dimensional reliefs and applied to the casket after firing the enamel. In almost abstract scenes, the pictures on the casket, which is only 29 cm long, tell the decisive episodes in the story of the particle of the cross contained inside.

The Shrine of St. Gislenus was assembled in 1865 of the enameled plates that remained of two reliquaries. The ten enameled plaques mounted on the shrine bear floral and abstract decorations that almost seem to show an Oriental influence. Two pit-enameled plaques with portrayals of the virtues are mounted on each side of the shrine. These plaques show most beautifully the strong expressiveness of the use of artistically effective gold surfaces as contrapuntal coloration to the manifold enamel colors. Pit enameling also allowed the portrayal of relatively small writing on such works of art. These plaques also show clearly how shadings of color and drawing perspectives can be placed side by side within one and the same pit without running into each other and with a three-dimensional effect. The medieval world found in the technique of pit enameling its own unique form of expression of deep inner piety as well as the fear of the hereafter, plus a portrayal of the sacred and divine world as something far away, at a vast distance.

Pit enameling was not produced exclusively in the Belgian Maas area, even though it was born in that region. Cologne, Hildesheim, Vienna, Rome, Limoges, Paris and Toulouse were all great centers of medieval art that had their own pit-enameling workshops. Not only were religious subjects portrayed in pit enamel, but so too were secular themes expressed in it, such as the famous *armillae* of the Emperor Barbarossa, which are preserved today as a great national treasure in the Germanic National Museum in Nu"rnberg. Surely it is scarcely possible for the present-day collector to obtain pit-enameled articles of the early and high Middle Ages on the market, unless auction houses offer individual enameled pieces—which may well disappear immediately into museums or private collections of wealthy collectors, unattainable for the average hobbyist. Still in all, it is worthwhile to deliberately seek out these treasures in the churches and great museums of Europe, to acquaint oneself with these remarkably beautiful and precious works of medieval European art.

Limoges Painter's Enamel

Without a doubt, painted enamels from Limoges are the epitome of any collection of enamel art. This applies above all to the early works of the 16th and 17th Centuries, which were never again equaled in their splendor, variety, and careful creation. Though Limoges already played a major role in the history of enamel in the 13th and 14th Centuries, it did so again in the 16th and 17th Centuries as the center of *painter's enamel*—so called because the enamel was applied to an intermediate layer of enamel as on a painted picture, without wires or any other disturbing metal barriers, and fired.

The chief characteristic of Limoges enamel, whether opaque or translucent, is that it could be applied so that colors did not run into each other, even without being separated by wires. Naturally this required several melting or firing processes, and pieces are known that can be safely presumed to have been fired up to twelve times.

This required a mastery of the technical difficulties involved and a precise knowledge of the differing melting points of different-colored enamels.

The second notable characteristic of painter's enamel is that, as opposed to cell enameling, the receiver of pit or deep-cut enamel—the copper plate—no longer played any role in the formation of the picture, since the metal was fully covered by the enamel.

The early painter's enamels of the 16th Century are regarded by present-day connoisseurs and collectors with great enthusiasm and known as the very incunnabula of the enamel art.

Due to lack of sufficient documents and archives, how the Limoges artists got the idea of producing enamel in this form is not known today. In any case, two pieces made in this manner have already been traced to the mid-15th Century; a plaque from the reliquary of Saint Sebastian (Saint Sulpice-les-Feuilles) and two small plaques in the museum at Poitiers, portraying a gentleman and a lady in colorful costumes. Dating these plaques is very difficult; on the basis of the costumes, one can determine that this type of fashion arose in 1450 and was no longer worn by 1500. Thus these two works of painter's enamel can be dated to this period. Yet on the basis of the development of painter's enamel, one can only guess. The French historian Labarte sees the basis of it in the epoch-making innovations in glass painting, also popular in Limoges at this time.

Until this time it was customary to cover individual small pieces of glass with a colorful coating and then put them together in mosaic form to make windows. At the end of the 15th Century, though, a change was made to painting whole portions of glass pictures on colorless glass panes in enamel colors and assembling them only in large pieces. The development of enamel painting runs absolutely parallel, for just as the lead fillets were reduced to a minimum in glass windows, the previously customary wire framings made by cuts in the metal were done away with in painter's enamel.

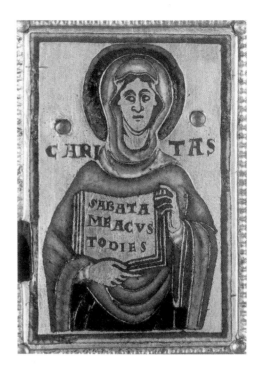

"Caritas"—pit enamel from circa 1150.

. . .without separating wires!

The New Glass Painting

29

Copies of Woodcuts

The role actually played by the Venetian influence on the workshops of Limoges remains undetermined to this day.

The influence of the woodcut, an art that attained it's peak at that time, on the Limoges workshops is clear, though. This is true to the extent that the Limoges artisans actually copied the woodcuts of great contemporary artists in enamel painting. The enameled plaque (see page 12) of Jean I. Pénicaud, already described, is clearly a copy of a Crucifixion scene by Albrecht Dürer. While the history of the Pénicaud workshop is well documented in the Limoges archives, it is questionable whether the often-named workshop of Monvaerni actually existed at all.

In the Limoges archives, fifteen enamelers are listed for the 16th Century, but none by the name of Monvaerni. Some forty well-known works have been ascribed to this artist—forty works whose actual authorship will probably remain undetermined.

Nardon Pénicaud

One of the most famous enamelers was Nardon Pénicaud, who must have run a large studio, for countless works of the early 16th Century have been ascribed to him. Originally, Nardon (a nickname for Leonard), who has been described as the ancestor of a whole dynasty of artists, was a vineyard owner in Limoges; he was later a consul. Just when he first devoted himself to the art of enameling cannot be determined from the archive data. Yet he created such works of unequaled beauty as a crucifixion picture that can be admired today in the Cluny Museum in Paris; it shows not only the usual New Testament figures but also a knight, a priest, and the coats of arms of France and of King René d'Anjou.

Albrecht Dürer: Crucifixion of Jesus with three angels, 1516 (Meder 182/Knappe 335).
During Dürer's lifetime his woodcuts already served as models for many applications to other artistic media. Shortly after Dürer's death (Nürnberg, 1528), all of Europe experienced a "Dürer renaissance." At the court of Rudolf II, Giovanni da Bologna cast entire sculpture groups in bronze. Dürer originals were already collected and sold for high prices in the 16th Century; it is thus no surprise that the enamelers in Limoges also worked from Dürer's models, including, for example, Nardon Pénicaud.

Nardon Pénicaud is also one of the late-medieval artists who signed their works; the inscription on this picture reads: Nardon Pénicaud de Limog(ia) fe(cit) p(rim)a a dié aprillis anno 1503."

It is regrettable that even these artists, as well as many others, did not create any original designs for their enamelwork, but rather worked from the plans of others. Toward the end of his career, Nardon Pénicaud preferred works by Martin Schongauer as his models. The enamelers of Limoges presumably regarded themselves as artisans of art rather than as independent artists.

What fascinates us so much about the Limoges enamel of the time is the intensive coloration that is so unique in this work. On the works of the already mentioned Jean I. Pénicaud we find a dark violet, hardly ever attained again, on faces, a lively, intense yellow and a no less intense purplish red.

Unfortunately, many enamel artists of the 15th and 16th Centuries are not known to us by name. Thus we have no choice but to name them according to their works, i.e., the "Master of Louis XI's Triptych" (this triptych is preserved today in the Victoria and Albert Museum in London).

A similar triptych is found in the Historical Museum at Orléans. Art historians call the creator of these pictorial tablets simply the "Master of the Triptych of Orléans."

Both works, consisting of main tablets and folding wings, represent a refined courtly type of enamel painting that leaves nothing to be desired in terms of nuances of color and high-quality artistic delicacy.

It is unnecessary to explain how the "Master of the High Foreheads" got his name. We find works of his in the Louvre and in Brussels; each is an unusually laborious work, rich in figures, of the lamentation of Christ.

It is also needless to say how the "Master of Anäis" got his name. In the Sixteenth Century he created an unknown number of wonderful portrayals of Anäis, eight of which are now in the Louvre, four in American private collections, and two were in the Pringsheim Collection in Munich; where these last now are is unknown.

While the "Master of the High Foreheads" worked from late Gothic copperplate engravings, it can be assumed that the "Master of Anäis" worked from woodcuts that were made in Strassburg in 1502.

In that year a certain Johann Grüninger published an edition of Virgil ordered by Sebastian Brant, illustrated by an unknown artist. It is remarkable to what size enameled plaques could be produced in one piece. The Vienna Museum of Art History owns a round plate of Grisaille enamel (see Enameling Techniques, page 13) made by Pierre Reymond (Limoges, ca. 1513-1584?), with a diameter of 45.2 cm. One can only guess at how many firing processes were necessary to create the fine shades and nuances in the drawing of Diana's holiday procession, the grotesque frieze, and the lively feminine portrait in monochromatic style on this splendid plate.

The terms "Limoges School" and "Limousin School" are often confused. Leonard Limousin was, of course, an enamel artist who lived in Limoges as a consul at the end of the 16th Century. The earliest known work of his dates back to 1534 and shows a portrait of Pope Clement VIII; it is preserved in the Louvre in Paris today.

The Pénicauds formed a whole dynasty of artists, of whom Jean I, II and III, Johannes Pénicaud Jr. and Nardon Pénicaud are the best known.

Some features of Limoges painter's enamel that are of interest in the history of art are the surface application of enamel to the medium of copper, the creation of excellent true-to-life portraits, and the breakthrough of Renaissance art, with its effect on all of European art.

After the days of Limoges enamel painting, artists had no new ideas for enamel work. One of the climaxes of enamel art was attained with Limoges enamel painting. For collectors and visitors today, a trip to Limoges may make them feel disappointed by the enameled art that is offered in that city as the heritage of a great tradition. Yet it is important and interesting to trace the forms of enamel art farther, to Baroque, Rococo and Empire artistic styles. In any case, all these eras only built on the ability of the enamel artists of the 16th Century and before.

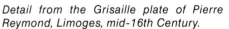

Dark Violet to Purplish Red

Detail from the Grisaille plate of Pierre Reymond, Limoges, mid-16th Century.

Medallion capsule with figures, enameled en ronde bosse; Paris, latter half of the 16th Century.

Baroque and Rococo

Enamel: The Art of Miniature Art

The end of Limoges enamel painting was also the end of an independent enamel art. Enameled works of art such as the Byzantine cell enamelwork, the high medieval pit enameling, and the painter's enamelwork in and after Limoges did not occur again in the history of art, even though enamel turned up again in many forms during the Baroque, Rococo, Empire and Biedermeier eras. Not in the form of independent works of art, of course, but as ornamentation, as added decorative elements...if you will, it was downgraded to mere decoration.

As early as 1543 one of the greatest sculptors of all time, Benvenuto Cellini, created a salt cellar of chased gold, partly enameled, mounted on an ebony base, for King Francis I of France. Later this piece passed into the possession of Archduke Ferdinand II of Tirol. This remarkably costly table utensil shows two human figures sitting face to face, enthroned on an ocean that in turn rises above the fluting of a base decorated with gold and enamel. The ocean waves and the clothing that swirls around the figures are made in the manner of deep-cut enamel and in part of émail repoussé. The colors of the enamel, blue, green, and red, give the decorative utensil an uncommon aesthetic charm. But here enamel is not the main point of the work of art, but rather serves as a decorative addition to a treat for the eyes. This early work is perhaps characteristic of the role of enamel from that time into the 19th Century.

Enamel in the Limoges style took up the task of portraiture on medallions, where it was framed by gold and precious stones. On caskets, tobacco boxes and other small containers enamel took over the role of an ornamental addition, in that garments, mythological figures or portraits were enhanced in terms of color with enamel.

Whole rows of stripes with moorish enamel patterns were worked into valuable teapots and other table utensils of onyx, gold or silver; these stripes were in turn decorated with pearls of precious stones.

House altars wreathed in pearls were made in very stylized form, including architectural elements in enamel applied by various techniques. The Munich court workshop was well-known for such work, which united translucent gold cell enamel, opaque enamel on copper, and émail repoussé for the outlines of figures in one and the same work.

The art of the Baroque and Rococo eras lacked nothing in terms of plentiful ideas and refinements.

The Moscow and St. Petersburg goldsmithing schools in particular created enamel works of art for the Tsar's household, works that more than enchant us today. The Moscow products were primarily bratinas (vessels with lids), vodka cups and spoons which, purely as a show of opulence, were made of pure gold and yet covered so thoroughly with precious stones, pearls, and enamel that scarcely any of the metal was visible.

Baroque, Rococo, Empire and Biedermeier shaped molten enamel in their own ways, according to their own styles. Clock faces and bases were decorated with the smallest enamel miniatures, so small that one might wonder how it was possible to apply the individual enamel colors. The luxurious life of the nobility knew no bounds. Table decorations were made of smoky or clear quartz, with gold plants growing on them, their leaves bearing translucent enamel on precious metal. Maria Theresia was given a breakfast service of gold, ebony, and porcelain, decorated with enamel painting, in 1750. Marie Antoinette obtained cut crystal table decorations that could be pushed back and forth on gold carts with enamel decor, made in a Paris workshop. The Archduke Karl Alexander of Lorraine received a surtout (table setting) that was 54.5 cm high, its crown of roses, chrysanthemums and carnations made of enameled metal, and its vessels decorated with enamel colors in the manner of Cantonese porcelain.

Enamel was often used as a frame for a valuable work of art. It was especially popular to mount ivory medallions in enameled frames of the finest filigree work.

The artist and goldsmith Jan Vermegen worked broad strips with pit enameling into the crown of Emperor Rudolf II in 1602.

In Italy, Commedia del'Arte figures were made in émail repoussé style; the costumes of Harlequin and Capitano glittered in colorful enamels over gold and silver on many table decorations.

Enamel could also be combined with mother-of-pearl, lacquer painting and bone to create optically enchanting ensembles. Toilette cases of such materials decorated the boudoirs of not only noblewomen, but surely of rich bourgeois women as well.

Agate bowls were mounted on colorfully enameled bases; the preference for oriental curiosities led to the making of covered vases and other small containers out of the stomach stones of Asiatic bezoar goats.

It is no wonder that this uncommon material was fitted with enameled lids in the Chinoiserie style that was coming into fashion.

Gemlike miniatures in which only the heads or figures were cut out of natural stone appear especially elegant to us today. The garments or hair decorations, on the other hand, were made of gold bands enameled so delicately that the whole piece achieves a unified effect of the highest artistic level. Bowls of matte green prase (a green semiprecious stone) were set with gilded bands on which airy Rococo designs were made with enamel.

The watchmaker's art used that of the enameler to set off its costly creations. The first flat watches by Jean-Antoine Lépine and the tourbillon watches of M. Breguet were valuable enough to be mounted in costly cases of enamel.

Workshops in London, Copenhagen, Munich, Ausgburg and Prague—in fact, goldsmiths in all towns and not just Europe—used enamel to give their works even more splendor. Enamelwork was surely made just for the nobility and rich bourgeoisie at this time. Collectors' pieces of this era, in a wide variety of quality, are available on the market now. Their prices are not based on age alone, but also on condition and quality, and of course, whether or not the pieces bear the (punched) signatures of famous workshops.

At the end of the 18th Century, medals changed from the older form of decorations (such as the Order of the Golden Fleece) to service medals. One of the first service medals to be awarded was the Austrian Order of Maria Theresia, founded by Maria Therasia in 1757.

In 1808 Emperor Franz I created the Imperial Austrian Order of Leopold. It is no wonder that medals, or the art of medals, used enamel at one time as an uncommonly decorative element, and that the collecting of medals forms a separate branch of collecting today.

Meanwhile, along with the 19th Century came the invention of new techniques of production as well as of enamel mixing. Thus today we find medals of collectable value with enamel in all colors, though the colors red, green, and blue are the most effective optically and thus most commonly used.

During the 18th and 19th Centuries, enamel was not merely a European means of decorating arts and crafts. Along with the Asiatic enamels, we know of many enamel works of that time from lands overseas, as many artists and artisans emigrated to lands such as America and Australia which offered them a better future.

Japanese Art and European Enameled Plaques

Enameled plaques, or more properly enameled placards, are desirable collectors' items with four to five-figure prices, whether they show the radiant Persil lady, the red "Erdal Frog" with the crown or the three Sarotti Moors that march glittering into the world. Attached to walls or doors, they advertised the products in question. When one hurries up the stairs of the London Underground toward daylight today, one often finds an enameled placard still unclaimed by collectors, exactly at the level of a stair, advertising a brand-name product that no longer exists. On fences, streetcars, and on the backs of park benches, enameled placards may still remind us that a clothes wringer belongs in every house or that insurance protects us from any threatening disaster.

19th Century Advertisement Graphics

These enameled placards are of historical significance in their relation to the development of advertising graphics toward the end of the 19th Century.

The decision to use single-colored enameled surfaces in graphic art was made by European and overseas artists under the influence of the Japanese art of the colored woodcut. It is thus not surprising that the first enameled placards appeared in Europe just at the time when the Japanese woodcut art was discovered.

In 1862 (London), 1876, 1878 and 1889 (Paris), Japan first made the woodcut art known to an astonished Western world at world's fairs. European painting was badly shaken by this exposure, and European and American commercial art was born.

This great influence is based essentially on the work of a single Japanese artist, Satsuki Haronubo (1725-1770); Japanese woodcut art owes to him the introduction of multicolor printing—some of his works were printed with ten or more plates.

What fascinated the rest of the world in this printing technique was the fact that one colored surface did not have to be modulated in itself but could have its effect as a single-colored surface within an ensemble of surfaces. And this very stylistic expression was used in the enameled placards made by the stencil technique.

What would Toulouse-Lautrec, with his fascinating placards of Montmartre dancers, be without the influence of the Japanese woodcut? What would all of commercial art, especially that of Art Nouveau Style, be without the art of Japan and thereby again Toulouse-Lautrec?

The art of the enameled placard, that had its heyday between 1900 and 1930 and is collected as art today, was based on that from the start. The stylistic characteristics, the use of a smooth surface, contrasting colors, concentration on essentials to the point of iconographic stereotypes such as using waves as a form of ornamentation, as seen in worldwide enameled placard art, can be traced directly back to the model set by Japanese woodcuts.

The enameled placards of Art Nouveau Style, Art Deco, and the era that lasted into the Forties come alive in terms of optical effect through their vigorous simplicity of form and contrast of intense colors. Perhaps they are much more expressive and effective as advertisements with this simplicity than the most refined modern types of placards and advertisements.

During the entire 19th Century, and especially toward the end, in the era that followed the Franco-Prussian War, many stylistic elements of previous centuries reappeared in forms of lesser quality. This mixture of styles, which turned up in enamelwork too (Neoclassicism, Neo-Baroque and Neo-Gothic), only gave way to new techniques with the advent of the Twentieth Century and its rejection of outworn traditions. The turn of the century brought the "new style" that we know as "Art Nouveau." Artistic rethinking became widespread, and enamel once again found a new form of expression.

Enamel: Art, Kitsch, Industry

Before the end of the 19th Century gave the world the long-scorned and now much-loved art of Art Nouveau with a deafening fanfare—which was subsequently adopted in, among others, the production centers of North America under the leadership of Tiffany—the world of art had many mixed-up resurrected styles. Neo-Gothic tried to recreate pictorial plaques in the Limoges manner but lacked the slightest originality of movement. The enameled pieces of this rediscovered Gothic style also suffer from a certain dullness, and its stylistic sensitivity was no longer able to harmonize with the basic religious outlook of the Gothic age. Neoclassicism and Neo-Baroque ended in what we call Historicism today. In the enamelwork of Historicism a single name can—indeed, must automatically—occur to us, that of the firm of Carl Fabergé, who operated a branch in Moscow as of 1887 in addition to those in London, Paris, St. Petersburg, Kiev and Odessa, and created the strangest and most abstruse jewels of all time for the Imperial house. The Fabergé firm's Easter eggs, made with gold cell enamel in opaque colors or translucent enamel on chased precious metals, are probably the most renowned. The Fabergé firm's tremendous production of Easter eggs reflects the religious significance of Easter in the Orthodox church. The Russian Imperial court's fondness for Easter eggs was, though, subject to unusual influences in terms of taste and a desire for unequaled luxury. Useful and decorative objects, even in unexpected areas, suddenly took the form of Easter eggs: bell-cord knobs, doorbells, table decorations, models of cathedrals, and containers for all manner of unnecessary ornaments all appeared in egg shapes. The Easter presents to the

On the basis of their technical production, the enamel "placards" utilize the formative medium of large surfaces, as European commercial artists had just learned to do from Japanese art (woodcuts).

Carl Fabergé

From the studio of Carl Fabergé: an Easter egg table clock, crowned by a bouquet of white Madonna lilies. The egg's base and body are covered with golden yellow translucent enamel. The clock numerals, formed of jewels, are on a white stripe of opaque enamel. The movement of this costly piece was made by Michael Perchin.

Empress are particularly extreme, from diamond-studded gold coaches to a massive gold portrayal of the Trans-Siberian Railway. Naturally, the House of Fabergé did not just make Easter eggs. Gold cigarette cases were covered with delicate enamel layers over chased backgrounds, and naturally set with valuable diamonds. The Tsar's gift of such a costly object with the Romanoff monogram was the equivalent of being decorated with a high Imperial medal—and they too were ornamented with enamel. But the religious objects made by Fabergé, such as cell-enameled icons decorated with pearls, did not equal the artistic, religious, and expressive power of the old icon art.

Though thousands of people rush to exhibitions of Fabergé art works, their admiration is directed more to the ingenuity of the objects than to their actual artistic content.

Fabergé achieved the ultimate in the field of enameling at the end of the 19th Century. Many goldsmiths tried to imitate Fabergé, but were unsuccessful.

A stark contrast to the luxurious decorative objects is created by the industrially produced enameled objects made since about 1860.

Enameled household goods of all kinds are found today, from the refrigerator to the stove. This industry developed very slowly since about the middle of the 19th Century. Utensils could be protected against rust and corrosion with enamel, and at the same time they were more hygienic because they were easier to keep clean; pots, ladles, towel racks and the like were covered with single-colored enamel. Following is a report on the beginning of the enamel industry:

"The enamel industry, which was regarded as an interesting but insignificant industry not too long ago, has awakened a heightened interest in ever-growing circles in the last 10 to 15 years, corresponding to its extraordinary development.

"An industry whose total production in Germany alone is valued at approximately sixty million marks per year, an industry that directly employs 20,000 workers in Germany and 12-14,000 workers in Austria-Hungary, has proved its necessity and its economic right to exist.

"The extraordinary development dates from the time when scientifically trained technicians and feeling replaced empiricism in the factory, researching and developing this hitherto closed-off, anxiously protected realm. This is a fact that our industry has in common with most others that are of any great significance. Though attempts were made to dispose of previously occurring, purely empirical mistakes in the past, it is the standpoint of the younger, scientifically trained generation of specialists that they work successfully not only to eliminate mistakes but to find their causes by studying the chemical and physical properties of enamels, and by obtaining the best possible raw materials and utilizing them in the best possible ways. And so it happened that enamels and enameled products, despite the constantly rising prices of raw materials such as feldspar, borax, tin, tin oxide, cobalt oxide, coal and sheet meta, plus workers' salaries, have not only been able to keep pace but to gain more and more popularity and acceptance of more types of use, thanks to simplification of production and improvement of work processes. This upswing that has been attained is not only attributable to the influences of modern enamel techniques, but to the fortunate conditions that have prevailed in all countries, where factory owners have self-sacrificingly encouraged all progress, have not avoided large expenditures, though some would prove to be unfruitful, and with many exemplary cases of goal-oriented activity, have had an inspiring and encouraging effect on the young technicians who, supported with a wealth of theory, ideals and good will, have dedicated themselves to this hitherto completely unknown profession, which lay far away from the path of the technician's typical activity. Challenging work, many problems, frequent stress, and strain awaited them. Thus the feeling of satisfaction was all the greater when a new advance was scored that gave new strength to manufacturing and opened new paths.

"It would amount to carrying coals to Newcastle if it were said that one and the same formula did not work for every business that used chemical compounds of the applicable materials, the sheet metal, the type of firing, the muffle and melting ovens and the working methods of enameling itself that are of definite influence on the utilization of an enamel recipe. The real professional will find it easy, wherever he is, to produce the suitable enamels for factory use within a short time. The differences in the recipe are all too often matters of very small but important differences, for essentially the structure of the recipe is the same everywhere, since the enamel represents a chemical compound that can be varied only within certain narrow boundaries. Thus the value and the ability are not in the recipe book but in the experience and intelligence.

"In addition, the experiences of the technician in the factory are not his exclusive property, since it has probably been gained thanks to his intelligence, to be sure, but almost always through the often expensive means of the factory owner."
(From: J. Grünwald: *Theory and Practice of the Sheet Metal and Molten Enamel Industry*, Leipzig 1908.)

From Art Nouveau to Post-Modern

Art Nouveau means avant-garde above all else, and avant-garde means being ahead, creating something new as opposed to the ordinary, free of confusing and disturbing perspectives of the day.

The very name of this era and style expresses it: the German word "Jugendstil", taking its name from the Munich journal "Jugend" (Youth, since 1896), the French expression *"style-nouveau"*, the term "Secession" still frequently used in Austria— which is, in fact, a concept from the history of the United States, referring to the attempted withdrawal of the southern states in 1861. Within art history, "Secession" means the separation of a group of artists from an existing artistic movement to form a new group, such as happened in Munich in 1892 under the leadership of Franz von Stuck, in Vienna in 1897 under the leadership of Gustav Klimt, and in Berlin in 1899 under the leadership of Max Liebermann.

What had happened? An originally quite intoxicating excitement of the 19th Century at the technical potential of industry and handicraft for achievement (the 1851 London Exposition!) was followed by deep concern, social pessimism and social criticism. At the same time, art seemed to be on a dead-end street with its imitation of earlier styles, and nobody seemed to know the way out. A symbol of a new development appeared when in 1887 the Vienna Art Academy chose a new star on the horizon of art as its professor of historical painting: Hans Makart, as the successor to Anselm Feuerbach. The one was a representative of Historicism, the other the excited representative of a new Belle Epoque, who produced Art Nouveau like an erupting volcano.

Meanwhile the unbelievable difference between great luxury on the one hand and the unattractiveness of consumer goods for the small people had grown too great. Concepts such as formation, elegance and beauty did not belong to everyday articles. The artists of the Art Nouveau Style also saw the overcoming of these differences as their mission. The utilization in handicrafts of floral decorative elements came about at roughly the same time when hikers and "Wandervögel" sought a way back to nature.

To try to describe the entire effect and the whole wealth of ideas of Art Nouveau, its background and its effect on the formative process, even in our time, would be going too far here. It may, though, be mentioned that an artist like Peter Behrens (Member of the United Workshops of Munich and later member of the Darmstadt Artists' Colony) became an artistic advisor for an industrial business like AEG in Berlin in 1907, for the first time in history, and created ground-breaking designs for various electric utensils. This constituted a decisive step in the commercial art of the turn of the century—from an artistic individual piece to factory-produced objective art for everyone! The inhuman chasm between machine production and creative, handmade preparation was thus bridged. The days of Art Nouveau left us with not only a number of valuable individual pieces but an even greater number of collectable mass-produced goods.

The spirit of the new lines, the sympathetic portrayal of plant patterns in ornamentation practically apart from the object, came naturally to enamel, which served splendidly in the essential sense of the word on account of its formability. Enamel played its own individual role in the jewelry and finely made utensils of Art Nouveau, because the artist was able to unite quality of form *and* beauty in a practical symbiosis. In addition, it enabled industry to produce enamels in the most wonderful color variations.

Adolf Böhm (Vienna), placard for the Secession exhibition of 1902.

Kolo Moser (Vienna): Ornamental casket (1905-06); the two enamel panels at the sides of the front show two idealized youths.

Great names of Art Nouveau

Silver butterflies with widespread enameled wings, enamel jewelry, enameled teapots, tableware with enamel decoration, enameled porcelain, enameled painted vases, enameled furniture attachments, enameled jewelry boxes, enameled necklaces, enameled plates, and enameled pictures of figures "en ronde de bosse" were not only of great expressive power but also imbued with the unique liveliness that pervades all of Art Nouveau.only a few of these artistic artisans, but their names will at least represent many others: Friedrich Adler (Laubheim); Charles Robert Ashbee (Kent), whose early clay forms, influenced by Japanese styles, had a great influence on German and Austrian handicrafts; Louis Aucoc (Paris); Jean Beck (Munich); Rupert Carabin (Paris); Ernest Cardeilhac (Paris); George Cartlidge (Newport, USA); Eugene Colonna (Paris); August and Atonin Daum (Nancy); Jules Des Bois (Paris); Christopher Dresser (Mulhouse); Jan Eisenloeffel (Amsterdam), whose enameled silver vases and pots revealed the architect's influence by their strict forms; Edmond Harry Elton (Summerset); Paul Folott (Paris); Emil Gallé (Nancy); Ernst-Moritz Geiger (Berlin); Robert Holubetz (Bohemia); Archibald Knox (?); Luce Limner (London); Adolph Loos (Vienna); Hermann Obrist (Munich); Richard Riemerschmid (Munich); Harald Slot-Möller (Copenhagen); Theodorus K. L. Sluytermann (Delft); Henri van der Velde (Zürich and New York); Philip Webb (London); Frans Zwollo (Amsterdam); and naturally Louis Comfort Tiffany (USA), a name that has become more widely known since Audrey Hepburn's fascinating role in "Breakfast at Tiffany's."

The art or the objects that these people have given the world do not consist merely of jewelry and unessential ornamentation, as was produced unnecessarily for certain layers of society at the end of the 19th Century, but also of well-formed, beautiful utensils for everyday and everyone.

Thus the Art Nouveau artists had also set themselves an educational task, consisting of the distribution of tasteful and beautiful objects for all; a certain artistic-educational effect was to be accomplished thereby.

For example, as of 1892 the glass artist Emil Gallé, in a program definitely in contrast to his valuable single creations, also worked on the creation of beautiful vases in industrial mass production. They were to be affordable to the general public. He did not create from economic considerations, though, but rather concentrated on social and educational aspects. Artistic handicraft objects were to be created that could correspond to a standard of "art for the people."

Interestingly enough, to stay in the realm of glass art, the European production of the vases in Art Nouveau that are so desired today was concentrated around Nancy, even though we must not forget Viennese workshops such as Lötz and the great Bohemian manufacturers. In the spirit of Art Nouveau that aimed at creating refined effects, the use of enamel played a major role in the production of glass too. The outer glass layers of etched and smoothed coated glasses were often covered one more time with enamel to achieve a particular effect. Daum, Lalique, Galleé, and Lelat, to name only a few, created flower vases with enamel overpainting of remarkable quality. The stylistic characteristics used on them refer back to historic elements. Imitations of Arabic script gave diamond-cut enameled cups and flasks an Oriental appearance. Japanese stylistic elements flowed into the patterns. Elements of Chinese- and Japanese-style color woodcuts found their use throughout. Art Nouveau developed into an aesthetically, almost overrefined, style and also created in the end—despite its demand for "art for the people"—single items of great value. Such enameled objects of those days in which glass and metal were united are especially charming: covered glass containers with enamel coatings on their silver lids, flasks with silver settings that were in turn coated with enamel, vases with enameled metal decorations.

Art Nouveau developed into the rather strict form of Art Deco.

For a long time, the objects of art and handicrafts of Art Nouveau were avoided by collectors. Today single objects are sold by auction houses at the highest prices. Mass-produced pieces can also be purchased today in the antique trade, at more modest prices.

Only today are we able to figure out the stylistic trends of the Thirties, Forties, and following decades and define their individual styles. Only now are publications available on the jewelry of the Thirties, in which enamel likewise played a large role.

There have been new developments in the art of enameling during the most recent decades, since technology and industry have made it possible for the handicrafter to attain noteworthy results in this area. The charm of such objects, made by untrained amateurs, should not be underestimated.

Artists of our days have taken up enamel again as an individual means of artistic creation, and are creating valuable jewelry and enameled utensils that will surely rank among the collectors' items of coming generations.

Early "Islam" flask with Arab-like enamel decor, by E. Gallé (1894). "Champignons des Bois" vase by Daum (Nancy), of 1907; both vases are decorated with enamel colors and overburned.

BIBLIOGRAPHY

1. **Reclams** Handbuch der Künstlerischen Techniken, Vol. 3, Stuttgart 1986.
2. "**Restauro**", International Journal for Coloring and Pianting Techniques, Restoration and Museum Questions, 89th Year, July 1983, Callwey, Munich.
3. Forman, W. & B., **Limousiner Email,** Prague 1962.
4. Kyri, H., Vol. 10, Bayer AG, Uerdingen.
5. Hasenohr, Curt, **Email,** Leipzig 1924.
6. Aldred, Cyril, **Die Juwelen der Pharaonen,** Munich-Vienna-Zürich 1872.
7. Bitterling, August, **Lehrling der Emailkunst,** Leipzig 1927.
8. Grünwald, Julius, **Theorie & Praxis der Blech-und Gussemail-Industrie.**
9. Bürger, Willy, **Abendländische Schmelz-arbeiten,** Berlin 1930.
10. Stöver, Ulla, **Email, Kunst aus dem Feuer,** Munich 1976.
11. Goepper, Roger, **Kunst und Kunsthandwerk Ostasiens,** Munich 1968.
12. Feddersen, Marin, **Chinesisches Kunst-gewerbe,** Munich 1958.
13. Yenyns & Watson, **Chinesische Kunst,** Vol. II, Zürich 1963.
14. Leger, Anton (Ed.), **Ornamenta Ecclesiae,** Vol. I & II, Cologne 1985.
15. Kunst-Historisches Museum Wien, **Führer durch die Sammlungen,** Vienna 1988.
16. Schmoll, Helga & J. A., **Nancy 1900, Jugendstil in Lothringen,** Mainz & Murnau 1980.
17. Herdin, Marcus, **Der japanische Holzschnitt und sein Einfluss auf die europäische Malerei,** in manuscript.

In Closing: Why Enamel?

We owe the art of enamel to the apparently boundless ingenuity of the human spirit in its search for new means of artistic expression, its gift of presentation, and its sense of aesthetics.

Man loves jewelry, adornment, decoration. But what is unique is that enamel was for a long time a handicraft that was applied in the service of other purposes before it became an independent expression of art in European enamel painting, lifting itself boldly above the task of supplying an aesthetic enhancement to gold and silver or serving as a filling material in golden cloisons, where it formed a colored surface in golden designs between technically refined wire frames.

In the world's enamelwork, art and technical skill have united in fortunate way; in Chinese and Japanese filigree cloisonné work, in the shrines and altars of medieval Europe, in Russian enameled icons just as in the lively, enchanting jewelry of the worldwide Art Nouvea Style, as well as in the modern works of present-day workshops on glass, copper, gold, and silver.

To the ancient Egyptians, enamel was obviously only a substitute for precious stones, yet they mastered the technique of producing it splendidly, as the hawk pectoral of the young pharaoh Tutankhamen proves.

The industries of the early-capitalistic Nineteenth Century discovered that enamel could also be "tamed" and made in quantity, and protected sensitive sheet metal and other metals with layers of enamel, making their products more hygienic and more pleasant to use. But these industries also had to meet their customers' desire for decoration, and employed hardworking women to hand-paint individual patterns on the mass-produced goods.

Enamel is supposed to be beautiful, to enchant and fascinate, to adorn, add value, and give splendor. It is admired by most everyone.

The initial fascination of the silky-smooth enameled surfaces was probably because they came, in the truest sense of the word, out of the fire. Even when cooled, enamel still retains something of the firing oven's glow—and sometimes retains it for hundreds, even thousands of years.

Its second fascination is that every enameled piece is an inimitable unique work; even when the work is deliberately repeated, every firing process nevertheless provides different results.

So it is no wonder that collecting enamel ranks among the more elite passions. It takes good luck and understanding, aesthetically trained eyes and a sense for recognizing the "right" piece on the market, evaluating it and categorizing it in terms of its time period and style.

Medallion with painter's
enamel over pit enamel

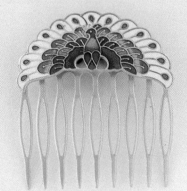

Decorative comb in classic
Cloisonné technique

Watch with painter's
enamel over pit
enamel

Winged scarab with Cloisonné
and window enamel

Owl pin in émail de
ronde bosse

Leda with the swan. Painter's enamel in
opaque colors on silver

Art-Deco brooch with translucent red
and opaque white enamel

Painted enamel
on metal, from
Persia

Painted enamel
on metal, from
Persia

Enamel from Asia

Above: Two Chinese enameled works in pit-enamel technique; at a glance, one could think this was Cloisonné work. The small 17th-Century incense vessel with a Fo dog shows enamels with strong color contrasts; the wide 18th-Century vase has a colorful dragon on a blue background.

Below: Two imaginary birds as candle or incense holders (18th Century).

The art of enameling reached Asia relatively late and was regarded there at first as being not fit to appear at court. Despised as "goods from the devil's land"—meaning Arabia—it must nevertheless have fascinated Asian artisans. For soon after enamel was introduced in China, the first enameling workshops were opened and produced enameled works of art of the highest quality.

Elephants as incense vessels (19th Century). Both pieces from Thailand are made in Cloisonné and have fascinating colors and technical quality.

Two Japanese enameled works, especially striking with their blue backgrounds. The decorations are made by Cloisonné technique and are pleasing to look at because of their attractive coloring and pleasant proportions.

Page 44: Upper left: The covered container (for ginger?) and the Cloisonné vase feature striking colors and technical perfection in their enameling. Lower left: The soft pink border on the upper part of this teapot is enchanting.

Page 45: The tall, bright-colored Cloisonné vase from China was surely used to hold fruit-tree blossoms. The bronze Japanese vase with its colorful enamel band offers perfectly proportioned enamel and monochrome bronze.

Enameled Boxes

Two boxes (probably originally part of a set of four) show, in émail paint of excellent quality, the Evangelists Matthew (with an angel) and John (with an eagle). The enameler succeded in capturing in enamel the artistic expressiveness of an oil painting.

Two tea chests in painter's enamel: the upper one shows a typical merchant shipping scene—perhaps a port of the East India Company—in nicely colored enamel; the lower one bears floral decoration in Chinese style on a strong yellow background.

Enameled Boxes

Two monochromatic cherubs with colorful robes play on a white enameled background; a scene in delicate painter's enamel. What is notable about these tobacco boxes is the fact that the underside of the lid was not covered in inexpensive counter-enamel, but shows a finely painted lady with a fan; perhaps she was—as was often the case—the giver of this valuable present to an admired gentleman.

A round lidded container with very finely painted outside and inside enamel, which shows fine cracks as a result of frequent use. The style of the romantic rural scenes suggests that it was made shortly before 1800.

A tobacco box with gold framing and—on a green background—idyllic bucolic scenes in fine, expressive painter's enamel with expert use of colors.

Enameled boxes with romantic scenes in painter's enamel and various decorations.

Enameled Boxes and Flasks

Cases and powder boxes, as accessories for women's handbags, are in themselves elegant and decorative. Here they are decorated with fine painter's enamel, i.e., translucent enamel over engraved silver.

The scene with the angel on the powder box at the top, and the cherub scenes on the heart-shaped and rectangular boxes indicate the end of the 19th Century as the time when these enameled boxes were made. All the pieces illustrated here express a joy in colorful enamel decor.

Erotic Designs on Enameled Cases

Below and right: Erotic pictures on cigarette cases and pill boxes indicate clearly that these pieces from the end of the 19th Century were made to be carried in the vest pockets of wealthy gentlemen; pieces of this type are the most favored of enamel and erotic collections alike.

A silver automatic music box with spring lid, showing Chillon Castle on Lake Geneva in émail paint. Art Nouveau.

Silver automatic music box with hinged lid. The lid shows a gallant shepherd scene in a bucolic landscape. Medallions in painter's enamel are set on the sides of the box.

From Automatic Music Boxes to an Art-Deco Vase

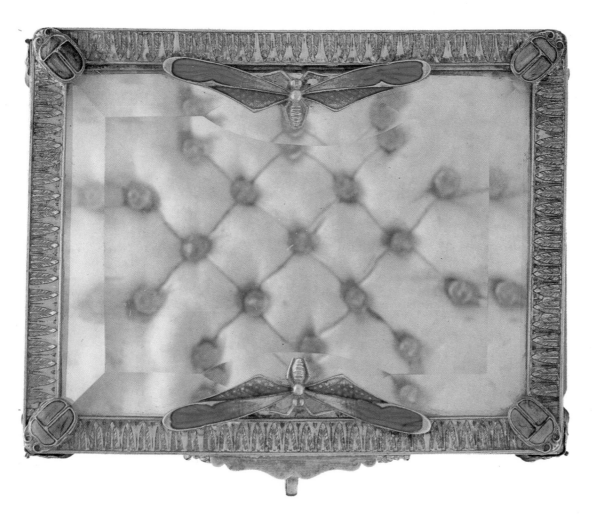

Translucent dragonflies and opaque scarabs in enamel decorate this jewelry box, which also has a wide enameled palmetto-pattern band. Probably turn of the century.

Large enameled surfaces blending into each other, in almost cubistic style decorate this unusual round vase from the Art Deco era.

Cigarette Boxes, Cigarette Holders and Pipes

Below: Two cigarette boxes and two holders in painter's enamel or émail en ronde bosse.

Right: Tobacco and opium pipes and a cigarette box—surely souvenirs of the Nineteenth Century from Persia. The Arabic style of the painter's enamel is based on traditional miniature painting.

Replicas from the Egyptian Museum in Cairo

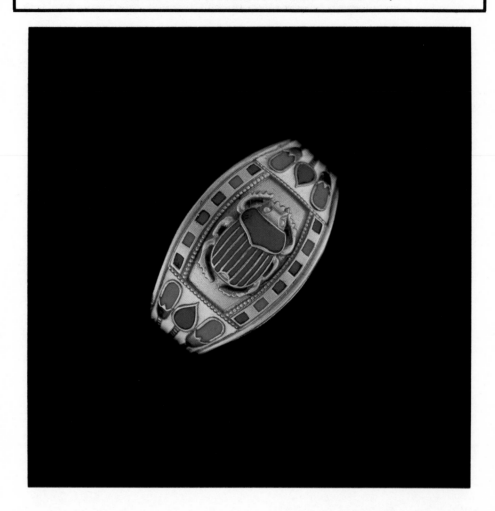

Museum replicas from the Egyptian Museum in Cairo portray jewelry from the treasure of Tutankhamen in Cloisonné technique.
Above: Bracelet with lotus blossoms and the sacred pharmacist (Scarabeus).
Opposite page, top center: Pectoral. Two falcons (Horus) crowned with the double crown of Egypt hold the king's name cartouche; outer left: The (Horus) falcon with the sun disc on its head spreads its wings protectively, forming the pectoral; outer right: Hawk (Nechbet) and snakes (Uraeus) of the Anch (life) symbol; bottom center: a worshipper—the king's cartouche.

There are "collectors' items" that are simply no longer available to the collector, as they have long since been in the possession of museums. Even if one of them were placed on the free market, the value would be beyond the means of even wealthy collectors. For the collector who values only originals, it is an excellent idea to collect quality museum replicas—a separate and appealing field for collectors.

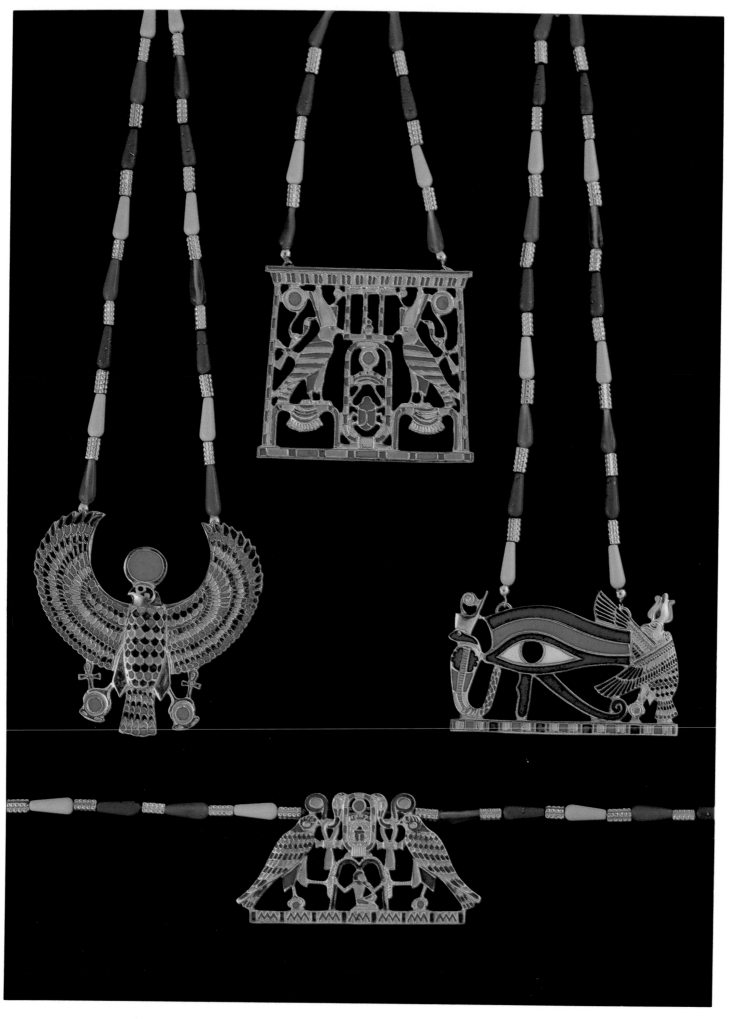

Enameled Souvenirs from Egypt

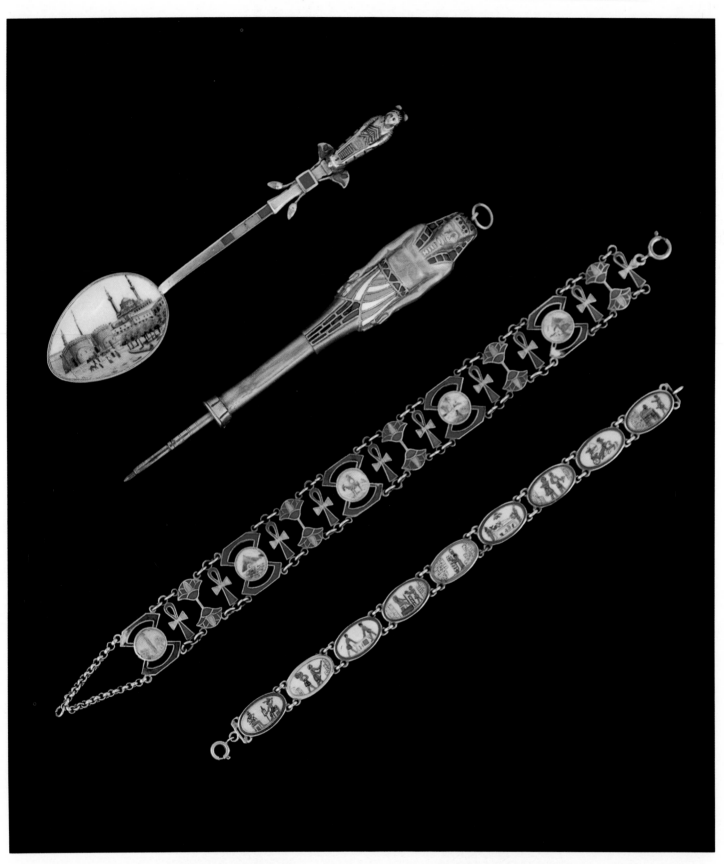

Souvenirs from Egypt. Above and right: Enamel decorations on spoon, stylus and bracelets—and museum replicas of a pharaoh's figure and a mummy case of precious metal and enamel.

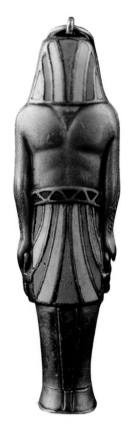

Enamel Painting on Jewelry

Enamel painting on jewelry—brooches, earrings, pendants—competed with miniature painting on ivory and achieved at least equal finesse, plus greater brilliance and considerably better durability. (The medallion at the bottom is shown again, greatly enlarged, on page 68.) Above: Pieces like these were often commissioned individually as portraits. For the collector it is particularly interesting to find pieces that go together like the brooch and earrings here.

Opposite page: Medallions, pendants, and brooches made by various techniques, also including different decorative elements such as pearls, precious stones, and gold chasing. The three pieces in Grisaille technique (second row middle, right and left of center) are especially unusual; the group of three dancers is on dark red enamel; the Erynnia head has a light brownish-blue touch; the portrayal of the pensive woman is set with precious stones.

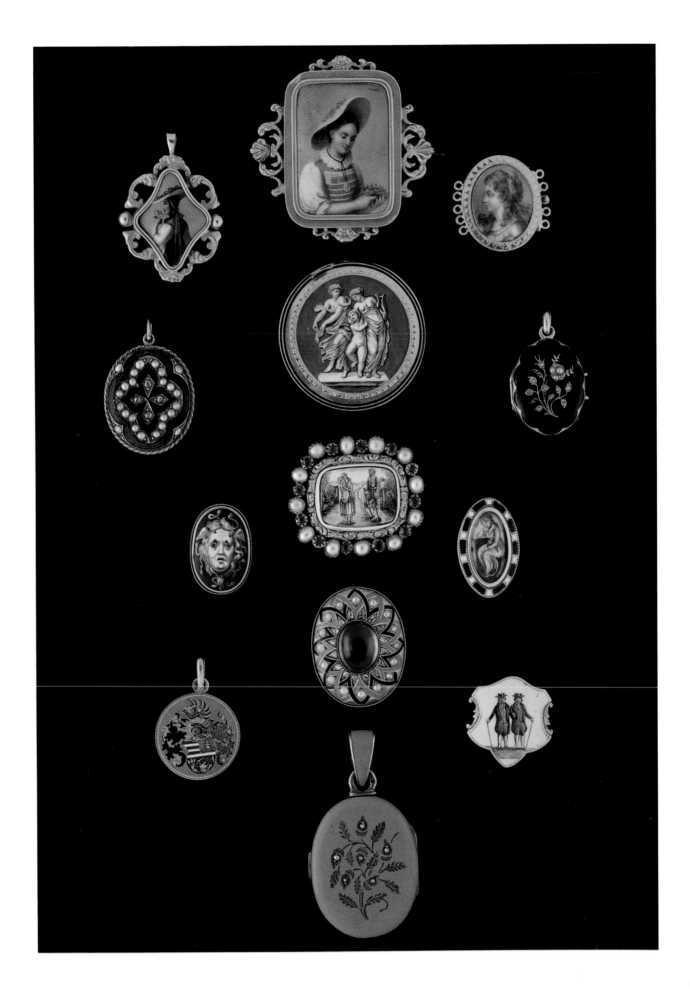

Jewelry in Various Enamel Techniques: Baroque to Empire

Below: Two brooches and pendants in finely made émail en ronde bosse or pit enameling. In these pieces, filigreed silversmithing work and glowing enamel colors blend to form especially charming visual impressions.

Below: This set of brooch, earrings and collar, in which pearls are cleverly arranged on black background enamel, is especially valuable.

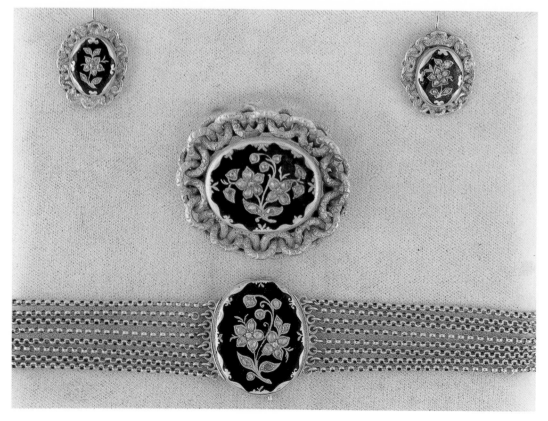

Two stickpins in particularly excellent painter's enamel; both the tender scene of mother and child and the Madonna scene with John the Baptist, based on Raphael's work, are made in artistically outstanding quality.

Jewelry in Various Enamel Techniques

Above: The filigree brooch from the early 19th Century shows a central medallion in fine gradations of Grisaille technique, probably portraying Apollo, or perhaps Athena.

Below: The lady with the fan and the diamond eyes could have stepped right out of the opera "Die Fliedermaus." The painter's enamel was fired in a pit in the medallion. The ear ornaments and the fan are set with pearls.

Art Nouveau brooches, pocket mirror, clock, and pin in various enamel techniques.

Three pendants from Art Nouveau days with fine painter's enamel, partially enhanced with Cloisonnés— the one at right with a touch of the erotic.

Jewelry in Various Enamel Techniques

Earrings in colorful Cloisonné enamel.

Baroque cameo surrounded by stones and blue enamel leaves.

Treasure-chest ring with deep blue enamel and gold decor.

Art Nouveau necklaces with enamel on silver.

Top center: Enlargement of a Limoges ring (1755; also at left). Below: jewelry in various enamel techniques; the necklace-pendant with amethysts and delicate "leaves" in changing translucent enamel creates an elegant filigree effect.

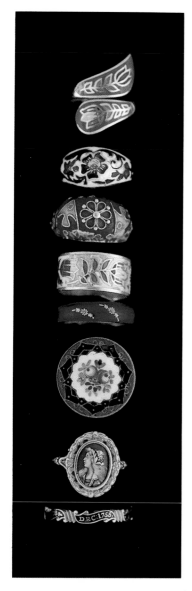

Enamel rings from various times; the ring from Limoges (see top center) is signed and shown above enlarged.

Enamel rings from various times; the Japanese ring (third from the bottom) shows fine enamel layers over three-dimensional precious metal.

71

Above: Painter's enamel with white background enamel; translucent enamels on silver, even in simple forms, reflect the faceted use of enamel on jewelry.

Opposite page: Opaque and translucent enamels on pendants and brooches. The splendidly colored peacock is made by Cloisonné technique. The shimmering blue translucent enamel on silver gives this metal a glowing quality.

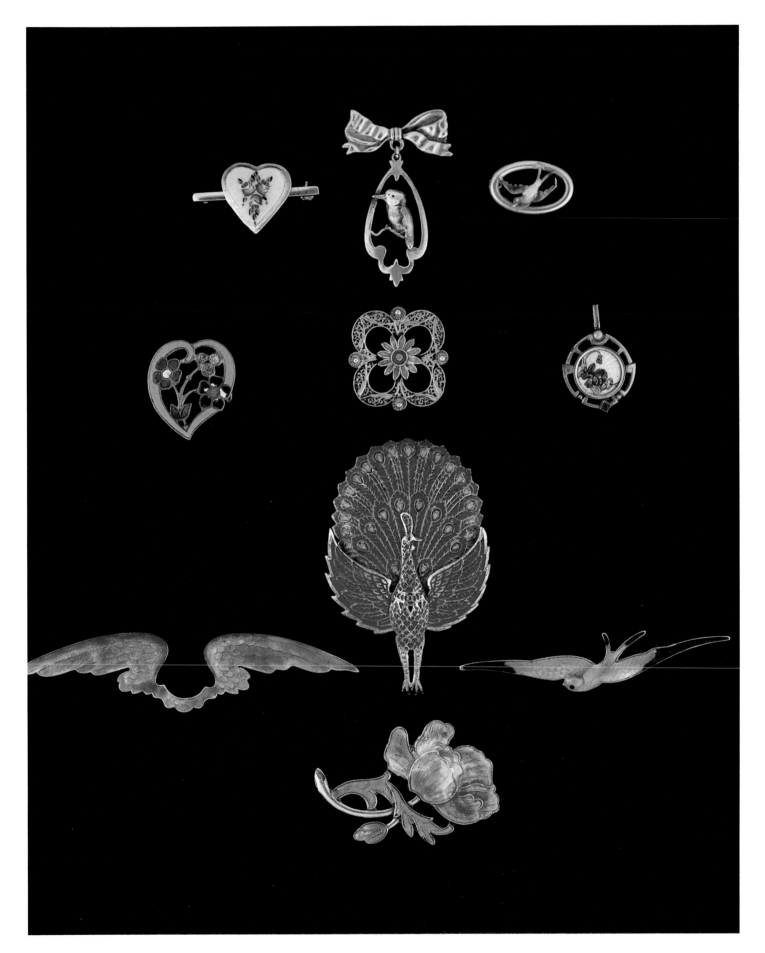

Above: Two silver pendants with portraits of women in painter's enamel. These pendants, clearly in Art Nouveau, show high-quality work in enamel by different methods.

Opposite page: Various pendants and earrings by different enameling techniques. Here it is particularly clear how elegantly enamel and precious metal can be blended.

Above and opposite page: These Art Nouveau and Art Deco pieces of jewelry gain a sense of fleeting movement from the elegance of enamel. The extent to which translucent enamels and synthetic stones blend to form elegant harmonies is especially striking.

Jewelry in Various Enamel Techniques

The brooch and the upper bracelet are silver enamelwork from Norway. Below: Silver enamel jewelry which shows what a variety of artistic expression can be attained with opaque and translucent enamel.

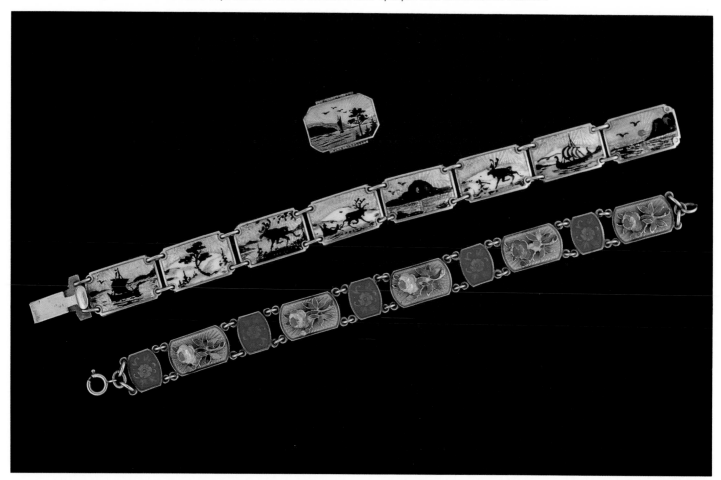

Above: Enamel jewelry; the bracelet in Fifties style and the brooch with its silver ornamental leaves are especially striking. Lower left: Silver enamel jewelry from Thailand. Lower right: Earrings and bracelets from Persia.

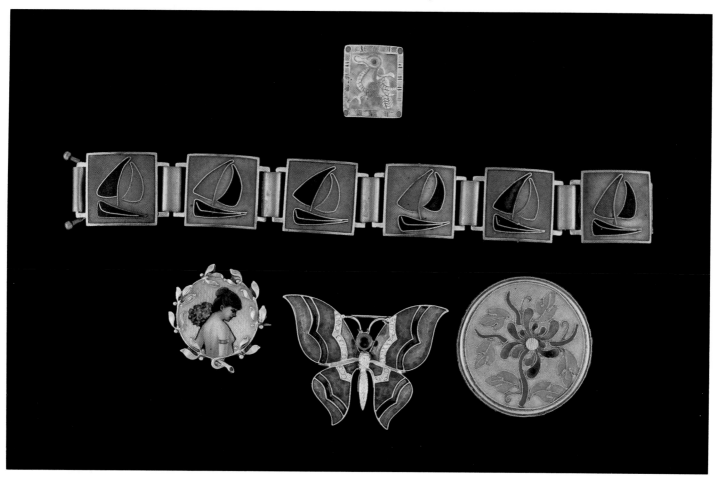

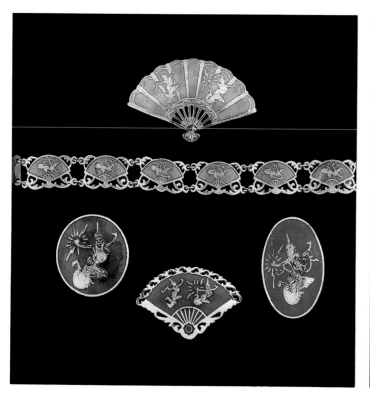

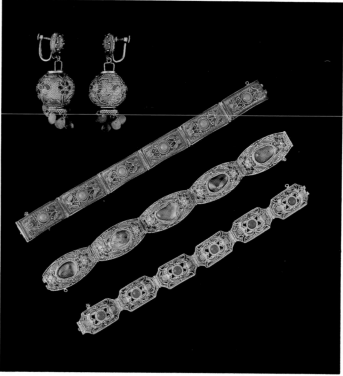

Brooches, Pendants, Children's Jewelry

Below and right: Jewelry is always collectible, especially when it is decorated with such elegant enamelwork as here; the pendants on the opposite page, probably originally intended as children's jewelry, are especially admirable.

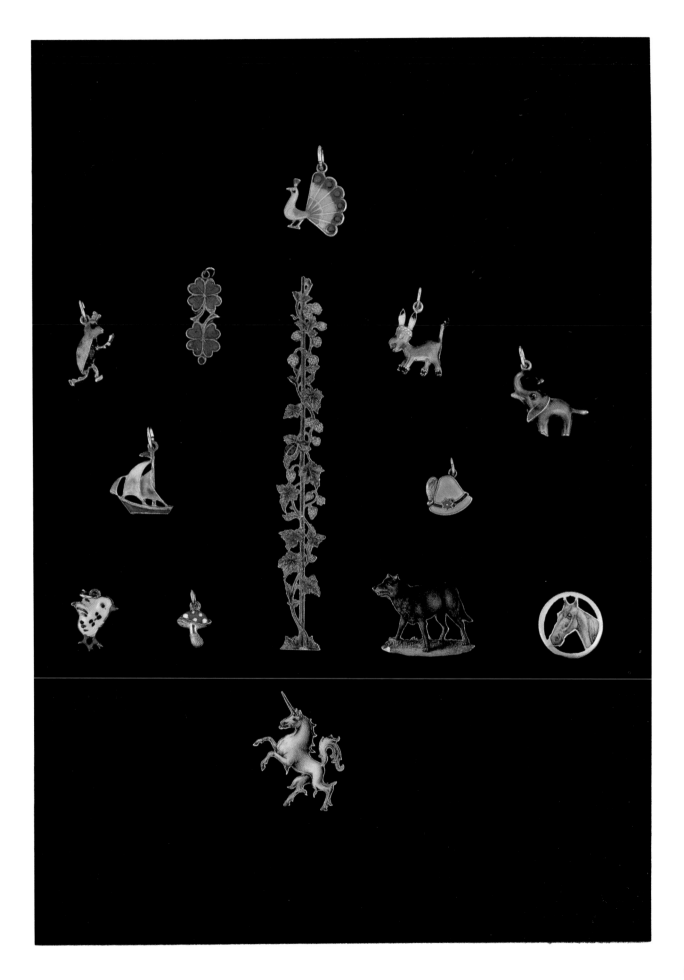

Pins in Butterfly Form

The elegance, transparency, brilliance, and color of the butterfly world have been transmitted and captured perfectly with "workshop" enamel and its techniques.

The playful moods of nature are expressed in the variety of the butterflies shown here. They give the collector information as well as excitement.

Art Nouveau and Art-Deco Jewelry

Below and opposite: Great sensitivity for color and shape, imagination and artistic creativity, coupled with technical ability and mastery of material make these pieces of jewelry stand out. Conceived as fashion jewelry, they nevertheless rank among valuable jewelry today. The wealth of variation in the materials and enamels gives these pieces striking elegance. (All pieces shown original size)

Painter's Enamel—Limoges

Religious art in enamel: Strongly expressive Grisaille enamel from Limoges, circa 1600. Fine gradations of color were achieved in the Grisaille itself, and the artist accented them even more with slight gold highlights, strengthening the effect further.

In the style of historic art from circa 1900, the enameler created this portrayal of St. George (underlaid with silver foil) killing the dragon. The figures and surroundings differ in terms of the delicacy of detail.

Religious Art and Cult Objects

Below: Enamel has a long tradition as a pictorial medium in the creation of religious objects. In these pendants—from the angel figure taken from the Sistine Madonna to the form of a Russian travel icon—the techniques change from painter's enamel to Cloisonné. Opposite: Collectors' items of museum quality (all approximately original size): In the small Renaissance quartz capsule (center) is mounted a portrayal of the Holy Family on their flight to Egypt, done in émail en ronde bosse. The body of the left cross features a red translucent enamel layer over silver—the (Bavarian) cross at right is done in country-style painter's enamel.

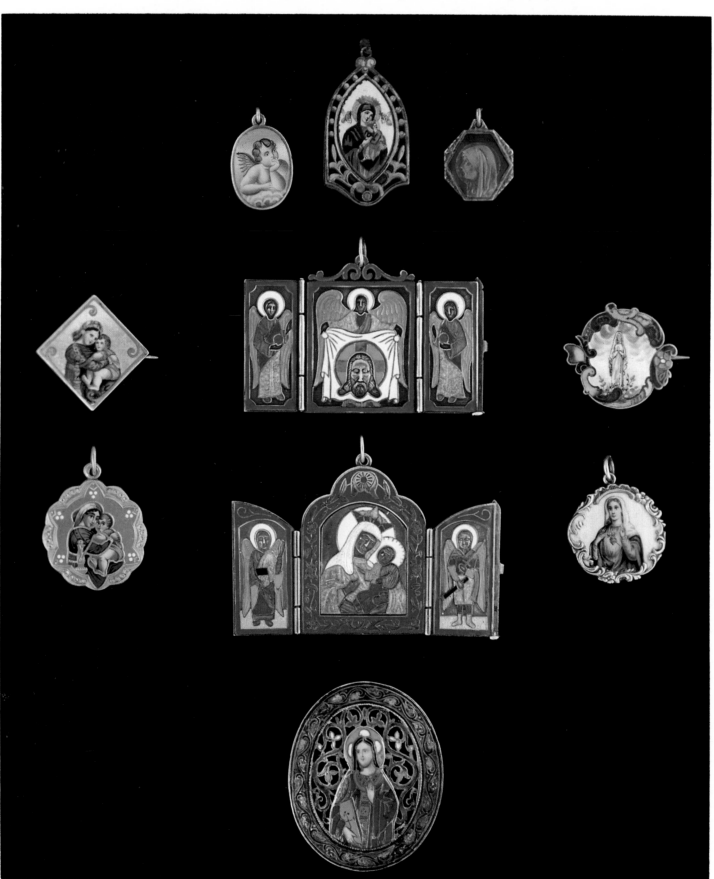

The Renaissance holy water font (late 16th Century) of cut quartz crystal is set in a frame decorated with translucent pit enamel.

Icons and Oclades

Collectors' items of museum quality: A Russian travel or table icon triptych from the early 18th Century—a Madonna at left and right, and in the center—according to the iconography—probably St. Nicholas.

As inspired portrayals of the saints, icons were recommended to the faithful for personal worship. Hence the costly decoration and partial surrounding of the wooden icons by golden Oclades. Yet the tender enamel decor could almost be called coarse. They take on their effect of lightness by imitating pit enameling, though they are made of raised cell enamel, since broad metal surfaces remain visible. The obvious signs of wear suggest frequent use as travel icons, probably also the cause of the pieces having broken out of the enamel. This damage adds to the charm and character of a piece. A typical example of a collector's item in which restoring the enamel would detract from the value and effect.

This small icon, probably Greek (though with an inscription in Latin letters in the silver Oklade), shows St. Nicholas. The bright-colored halo of colorful cell enamel, with corded silver wires, has the effect of a helmet.

Picture of a Jewish high priest in the form of a kiss-plate (as used in Catholic liturgy). This rare piece, 12 cm high overall, shows artistic talent in its painter's enamel on copper, framed by a painted border, set in a pearl-studded metal frame. A rare collector's item, surely of East European origin.

Icons and Oclades

Left and right: Two Russian icons, probably of the 18th Century. Since icons were always more than portraits and were meant to suggest the presence of the ones they portrayed, they were particularly respected.
Thus the inspiring effect of the rich adornment, here consisting of silver, fore-gilded Oclades, probably of the 19th Century if not from the beginning of the 20th, with decorative cell enamel, especially in light sky blue. The delicacy of the Oclades and the colors of the enamel take away some of the flatness of the Madonna and Christ Pantocrator.

A particularly interesting collector's item: a large drinking cup in raised Cloisonné with corded cell wires, made by Perchin, a master who worked for Fabergé.

A covered container with typically Russian decor in technically perfect Cloisonné.

Elegant collectors' items from the Fabergé studio. On the letter opener as well as the cuff links, the enamel provides a background for the actual decoration.

Above: Cigarette case. Below: Card case, each showing front and back. For the cigarette cases, silver wire was used to form the cells; for the card cases, corded brass wire. The effect of the boxes lies in their lavish decorations and smooth surfaces of green or white background enamel.

To understand fully the significance of the enameled Easter egg in Russia, one must know the importance of the Easter holiday in the Russian Orthodox liturgy and thus its meaning for the Russian people. The Easter holiday united heathen rites of spring with the religious awareness that faith and nature conquer darkness.

In the enameled Easter eggs from the House of Fabergé shown on these pages, it is especially striking that the enamels were fired on the background at different heights, that fire-gilded corded wires were used to form the cells, and that the enamelers also used the technique of painter's enamel within the individual cloisons. Thus these splendid creations took on a special, valuable effect.

Russian Upper-Bourgeois Enamel

Russian enamel; its decor suggesting the era shortly before 1900. It is notable—and especially typical of Russian enamel—that in all pieces corded wires were used to form the cells. Although these pieces were basically made by cell enameling, techniques of painter's enamel were used within the individual cloisons for decorative enhancement.

Religious-sect chalices with bowls of jade and bases with charming enamel decoration (Russian).

Complete sets like these are especially valuable to the collector.

Dessert and Vodka Spoons

Below and opposite: Russian vodka spoons, or fruit and serving spoons, in splendidly colored, almost Arabesque floral decor. Such pieces, even when first made, were intended more for prestige and show than for actual use. Long regarded as more or less kitsch, they are very desirable additions to any enamel collection today and turn up often in undamaged condition. Two features of these pieces are noteworthy: the all-around enameling of the handles and the use of inherently decorative or cordlike metal wires between the enamel cells.

The two spoons at the outer left and right show traditional designs, but in modern styles; the small spoon in the middle and that to its right clearly show the influence of Art Nouveau.

Enameled Tableware and Spoons

Finding sets of matching utensils like these is particularly fortunate for a collector. The floral patterns in Cloisonné technique, with corded wires, suggest the era of Art Nouveau.

Spoons like these were probably made for show rather than everyday use; otherwise they would not be in faultless condition. The fine enamelwork makes them valuable collectors' items. They are shown approximately life-size; the souvenir spoon from Lindau, with its miniature size, is especially rare.

Coats of Arms and Souvenir Spoons

Below and opposite: High-quality coat-of-arms spoons can be found today only in the antique market. Charming painter's enamel decorates the bowls of the spoons, with Cloisonné enamel on the handles identifying these rarities from various lands.

Objects of Everyday Use

Enamel as charming decoration for objects of daily use: a flask with Cloisonné enamel, silver comb case with painter's enamel, ball-point pen with translucent enamel over chased silver, signet with émail en ronde bosse and Cloisonné, fork with Cloisonné, spoon with painter's enamel, umbrella handle.

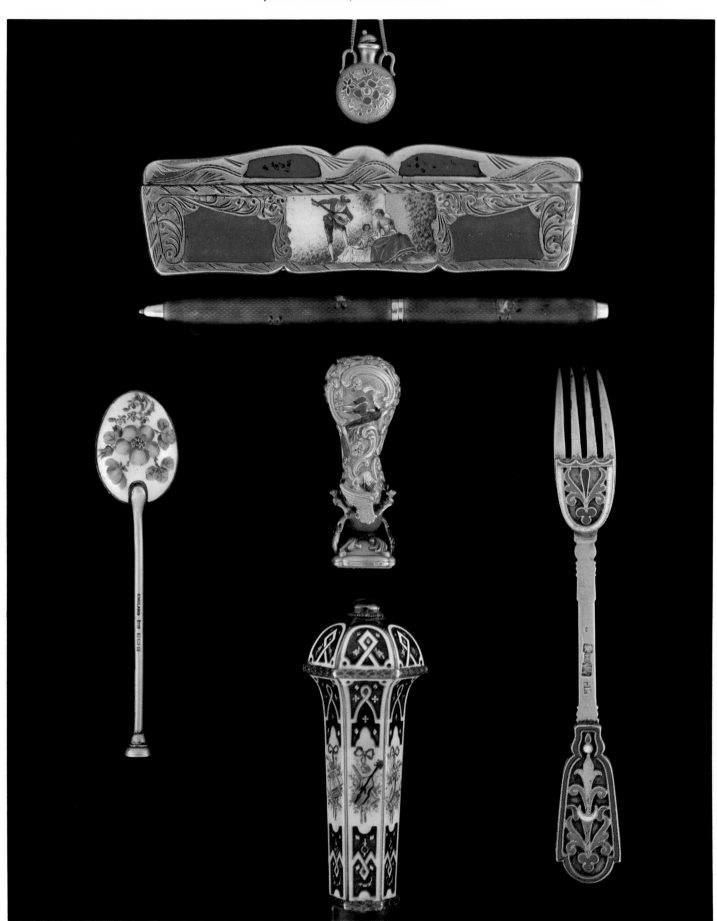

Pleasure items: lipstick case with painter's enamel on a white background, loupe in Cloisonné technique, bookmarks, lorgnon with pit enameling, whistle with country scene in painter's enamel.

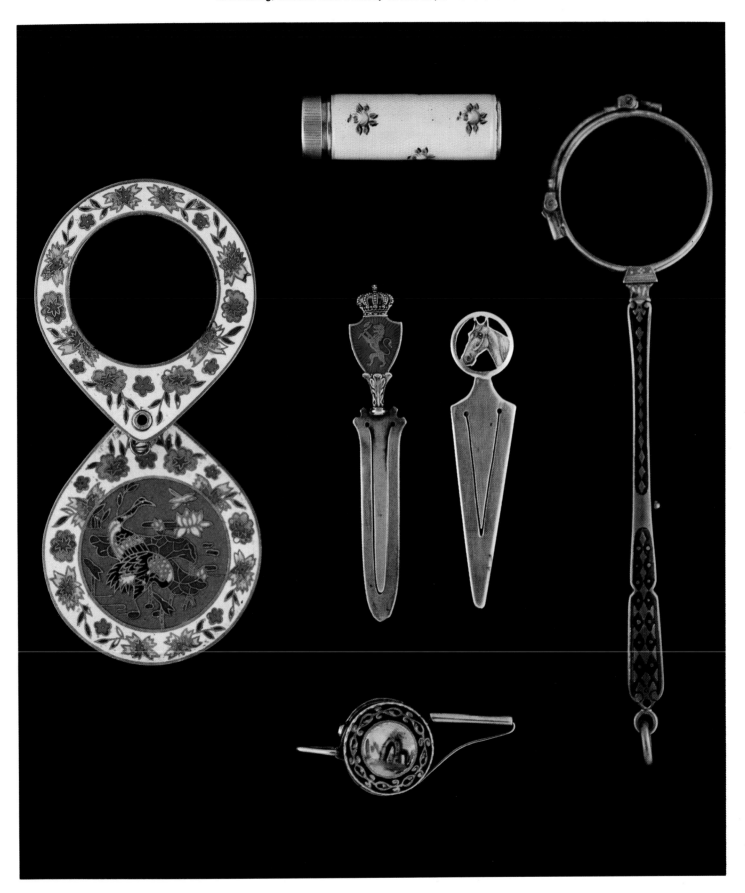

Opera Glasses

Opera glasses—as a status symbol at social events—are natural subjects for decoration and ornamentation. As collectors' items they are all the more valuable when they are decorated with beautiful enamel, as here. The successful combination of enamel and synthetic stones on three pieces is striking.

Belt Buckles

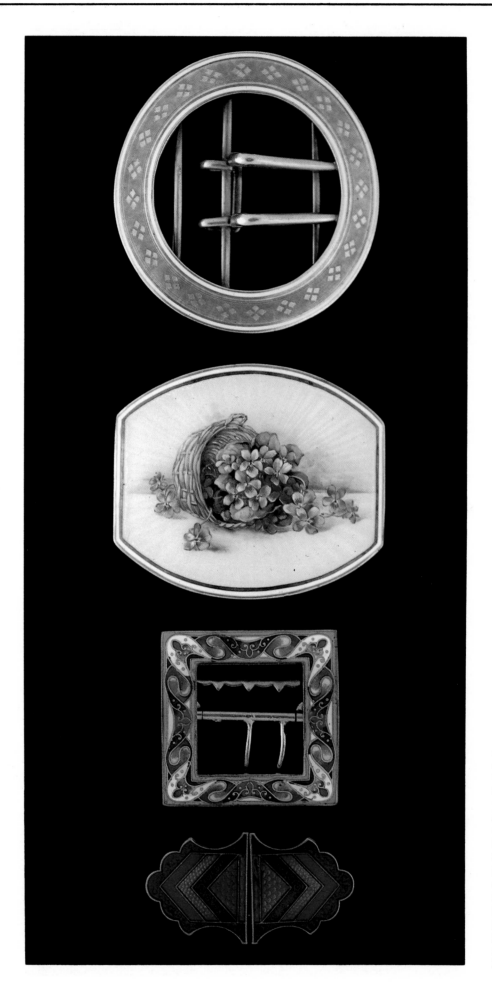

Belt buckles: from top to bottom, translucent Enamel on silver, delicate enamel painting, two pieces with cell enamel.

At right:
Belt buckles from various eras: at left, three belt buckles with almost Arabesque decor in Cloisonné technique, varying the lily shape, with floral Art Nouveau decor. At right, translucent enamel on silver, delicate painter's enamel, painted rose decor on white background enamel.

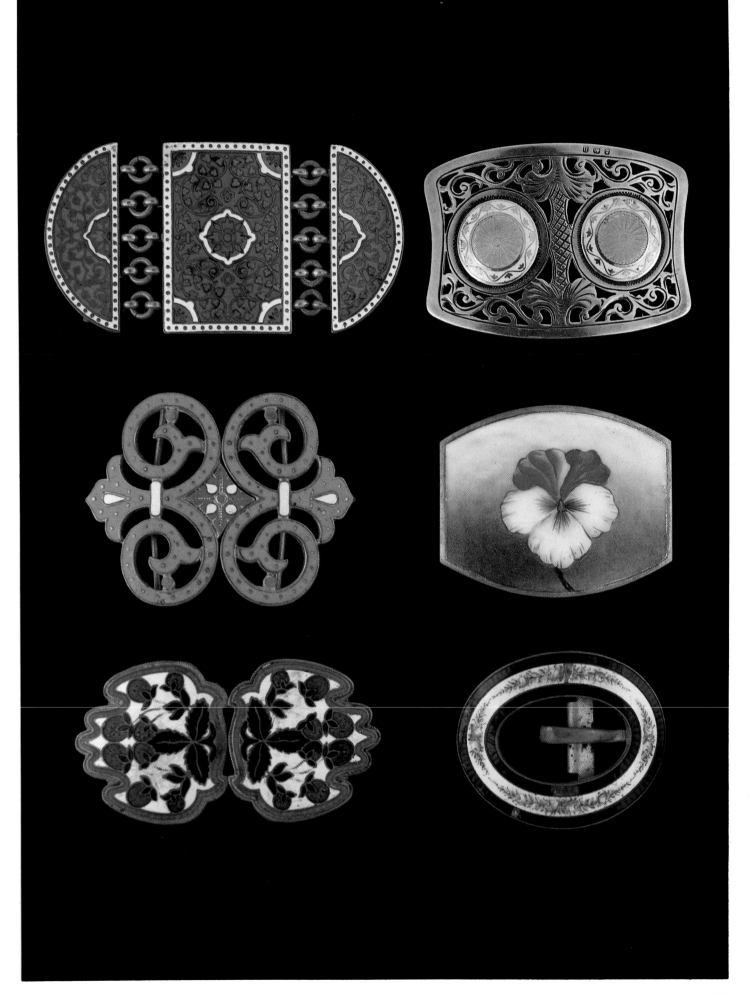

Two Art Nouveau ink bottles; the upper one with an enameled metal rim mounted on an onyx plate. Both pieces in Cloisonné technique, with splendid color effects, can be the main attraction of an enamel collection.

Small pocket mirrors with enchanting painter's enamel or (upper middle and right) enamel over chased silver backgrounds.

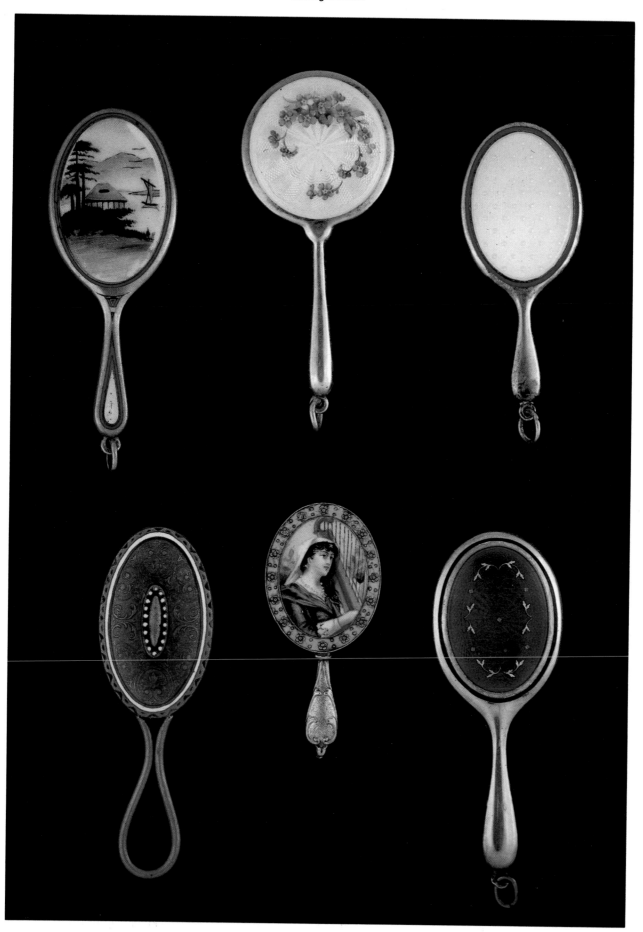

Small Plaques and Souvenirs

Below: Plaques of this type were generally made by the cell-enameling process.
Right: "beggar's bracelet" and key chain with local coats of arms, popular souvenirs in Cloisonné technique. The Tyrolean eagle in handsome dark red, large and small types with Edelweiss. These varying souvenirs were made by the painter's enamel technique.

A valuable clock on an intricate stand. Various enameling techniques were used to make this piece: painter's enamel on the clock cover and the round medallions on the base of the stand; pit enamel under the elephant and on his saddle blanket; a particularly elegant pit enameling surrounds the face of the clock; the angels' heads on the stand were made in the manner of émail en ronde bosse.

A rare pictorial clock in a fire-gilded frame, surely dating from the first half of the 19th Century. The round clock is set in an arched, oval, enameled plate, showing a Swiss scene in typical generic painting, as was very customary for decorations at that time.

Clocks and Watches

Left: Fine Grisaille enamel on a fire-gilded pocket watch. The portrayal of an idealized woman dates the piece clearly as from the beginning of the 19th Century.
Below: Original enamel decor on a farmer's pocket watch; the rural scene reflects a pleasant atmosphere.

Enamel Decor on Watches

As early as 1630, clock and watch dials and cases were being enameled in Blois, France. Since the gold of the cases was usually very thin, it could not stand the high temperatures needed for enameling on one side, and thus had to be enameled on the other side too (counter-enamel). In 1630 the goldsmith Jean Toutin of Châteaudun developed colors that could be applied very thinly and bonded permanently to the enamel when fired. Thus he invented a new technique of enamel painting. Scenes from mythology were favorite subjects when the Huaud brothers, among others in French-speaking Switzerland, produced watches with enamel of the highest quality from 1670 on, making use of Touton's technique. These watches were usually signed along the edge; they still turn up at auctions occasionally today and are highly desirable collectors' items.

Back in western Switzerland, where outstanding artists were active in and around Geneva in the 18th Century, Jean-Louis Richter made himself a lasting name by his "painting under enamel." When the background was artistically painted, it was coated with translucent enamel and fired several times. Thus depth and brilliance were created, with fascinating results.

Enameling watches has continued successfully into our own times, with all manner of subjects used.

Clocks for noble and upper-bourgeois households were not only instruments of time but also status symbols on the basis of their value; these clocks were decorated in valuable Cloisonné enamel. The pocket-watch holder at the lower right was made for a watch to hang in whenever it was not being worn.

Enamel on Valuable Watches

Enameled watch chains with Swiss symbols. At left, a chatelaine for a pocket watch; above, a watch with formed chains and decorations plus painted Edelweiss; below, an Alpine pocket watch; right, a green enameled watch key; below, an enameled Art Deco wristwatch; below it, an enameled pocket watch for a lady. Opposite: Modern-day enameled table clocks from China.

Above: Enameled sheet-metal containers made for children, with their own nostalgic charm.

Below: Enameled sheet-metal utensils for the 19th-Century doll house.

This Japanese teapot showed enamel-painted decor of naive yet beautiful charm.

Commercial Enameled Goods for Doll Houses

The area of antique-doll collecting has developed tremendously. Doll shows and sales, as well as specialty shops, have afforded a close look at the secrets of the doll world. A new branch of industry enlivens the scene of the world's oldest toy, and accessories are no exception. Beloved and desired above all are the charming furnishings of doll kitchens and houses and those that are suited to decorate the scene for dolls in living rooms and shop windows and bring "life" into that scene. Plates, bowls, pitchers and many other utensils have thus won a favored place in terms of popularity among collectors.

Child's plate and sheet-metal cup with stencil-painted enamel in strong, full colors; unforgettable utensils from childhood.

Commercial Enamel—Mass-Produced Goods

Spoon rack with ladle and sieve spoon—at right, a water dispenser.

Commercial Enamel—Mass-Produced Goods

Who does not know and love them, the pleasant plates, cups, pots and pitchers of Grandma's day? Who does not feel transported back to his youth at the sight of them? These beloved utensils are usually no longer present in our own households. They had to give way to porcelain and other up-to-date objects. All the more pleasant that one can find and buy so many of these things at antique and flea

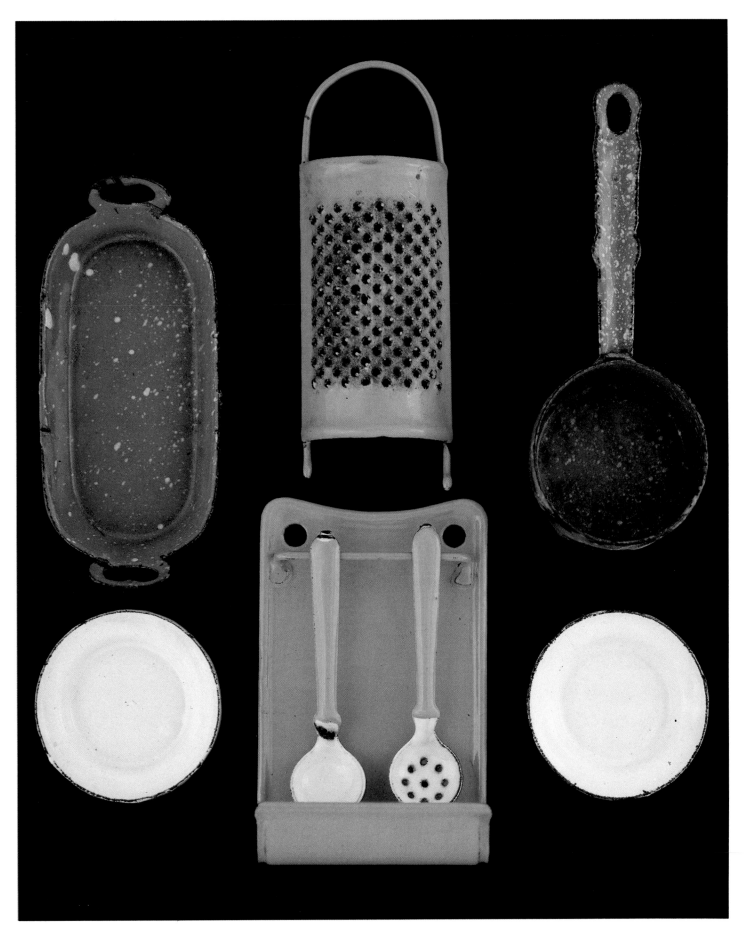

Enameled signs are a separate field of collectors' items today and sell for high prices at auctions. The enameled placards on these two pages date from the early days of our century and show various stylistic characteristics. When one considers that every color of enamel had to be fired separately, one's respect grows for the ability and industrial preparation of these durable advertising signs. To make such placards, the individual enamel layers were either sprayed on with stencils or sieved onto the background through stencils. In any case, the applied enamel had to be dried before the whole piece could be fired again. Surely the cost of producing enamel placards must have been rather high.

Since the enameled placard could be damaged easily on account of its size and the sensitivity of the material, these signs are quite rare and highly prized as collectors' items today.

Medals and Honors

Left: Order of the Lion of Zähringen, "Knight's Cross First Class in Gold"; top, "Star of the Special Level" with genuine gems in the middle; below, Order of the Crown of Thailand, with sash pin and star below; right, "Pour le Merite" (for service), former high Prussian honor.

Mardi Gras Medals

Mardi Gras or Carnival medals are as difficult to earn as the more "serious" medals. Here too, the quality of the cell or painter's enamel is striking.

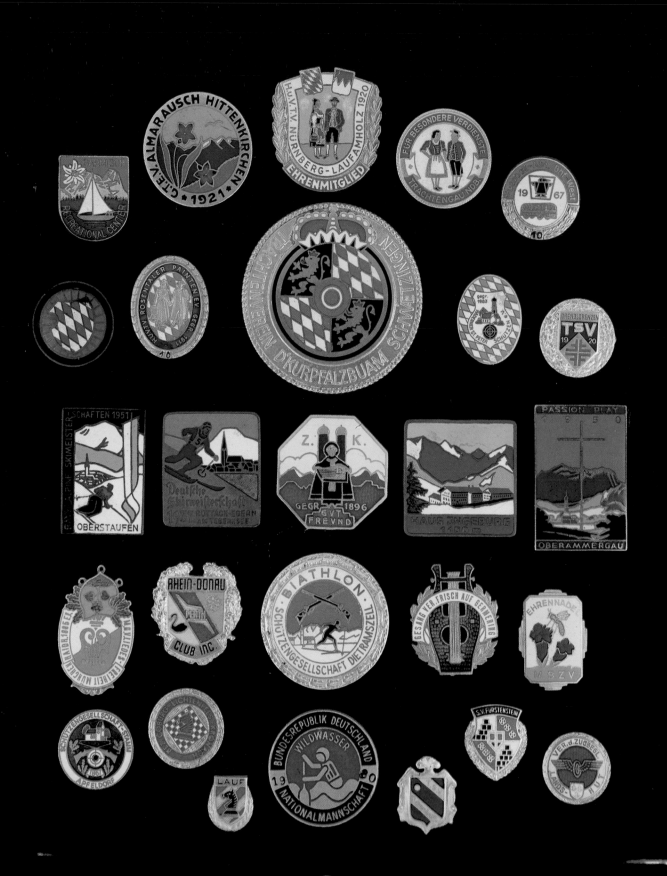

Left and below: Low-priced pins and plain emblems have increased the value of enamel badges and club pins and turned them into a fascinating and extensive area for collectors.

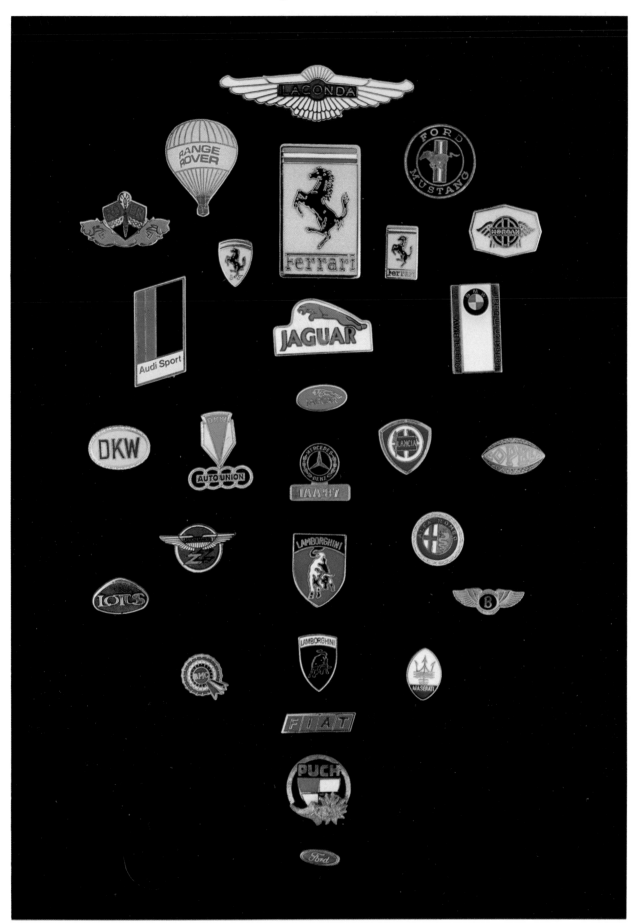

Enamel of the Fifties

Enamel in the style of the Fifties.
Right: The enameler has developed purely pictorial elements on the lid of the cigarette box and the surface of the ashtray by applying color and firing correctly.

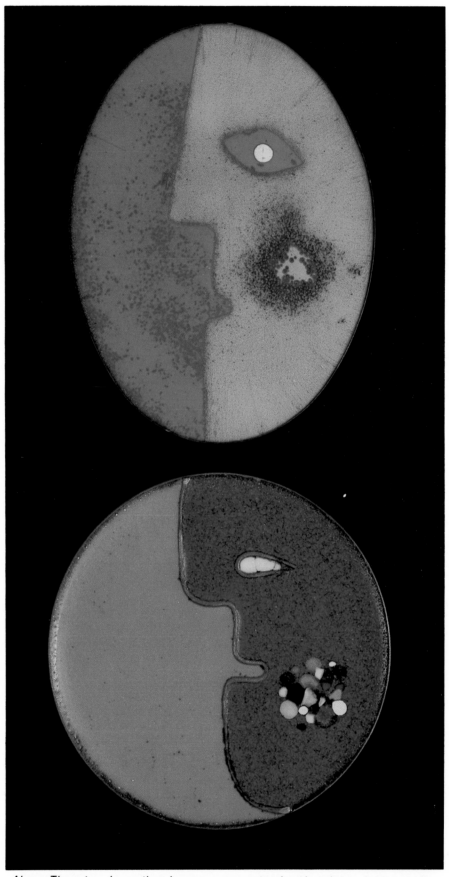

Above: These two decorative pieces on copper stand out in color contrast as well as clear and meaningful lines.

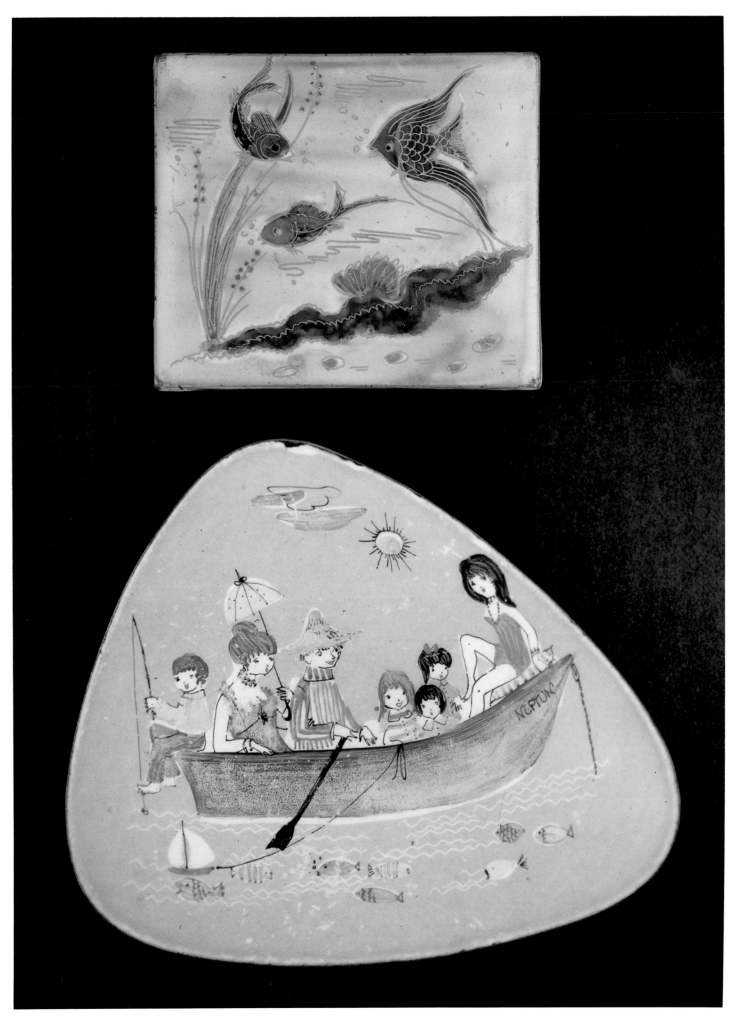

Enamel Easter Eggs and how they are made

The demonstration set from China shown above makes clear what work processes are needed to produce an enameled Easter egg. Box 1 shows the raw egg of copper.

Boxes 2 and 3 show the conditions after applying the cell wires and after the first application of translucent background enamel to the raw egg.

Boxes 4 and 5 show how the colors are applied to the cells and fired, one after another.

In box 5 the cells have already been filled with enamel to the height of the wires. It is easy to see the dark rims along the cell wires—the result of the firing process.

Boxes 6 and 7 show clearly how often collectors' items like these must be refired and repolished until the enamel has its desired color effect and the cell wires remain as gold lines between the cells of enamel.

Opposite page: Modern-day enamel Easter eggs from China, in varying sizes, with special emphasis placed on harmonious color effects between the colored background and the decor.

Enamel Easter Eggs

Three Chinese Easter eggs in vividly-colored, contrasting enamel. In filling the cells, the enameler was able to apply changing colors even within cells. The fact that the whole enamel application stands out from the background enamel like a high relief on these pieces is striking.

Containers, tea bells and thimbles, likewise of vividly-colored enamel that stands out over the shimmering gold metal surface (China).

Enameled Animal Figures from China

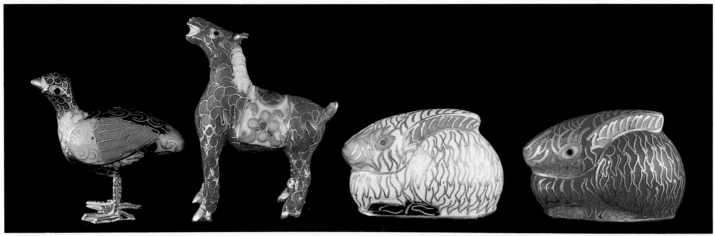

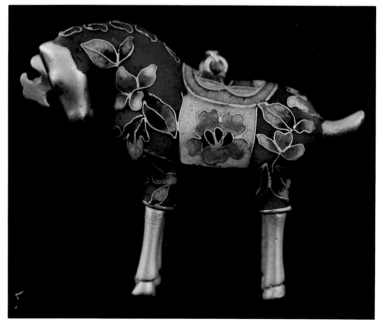

Lighthearted Objects

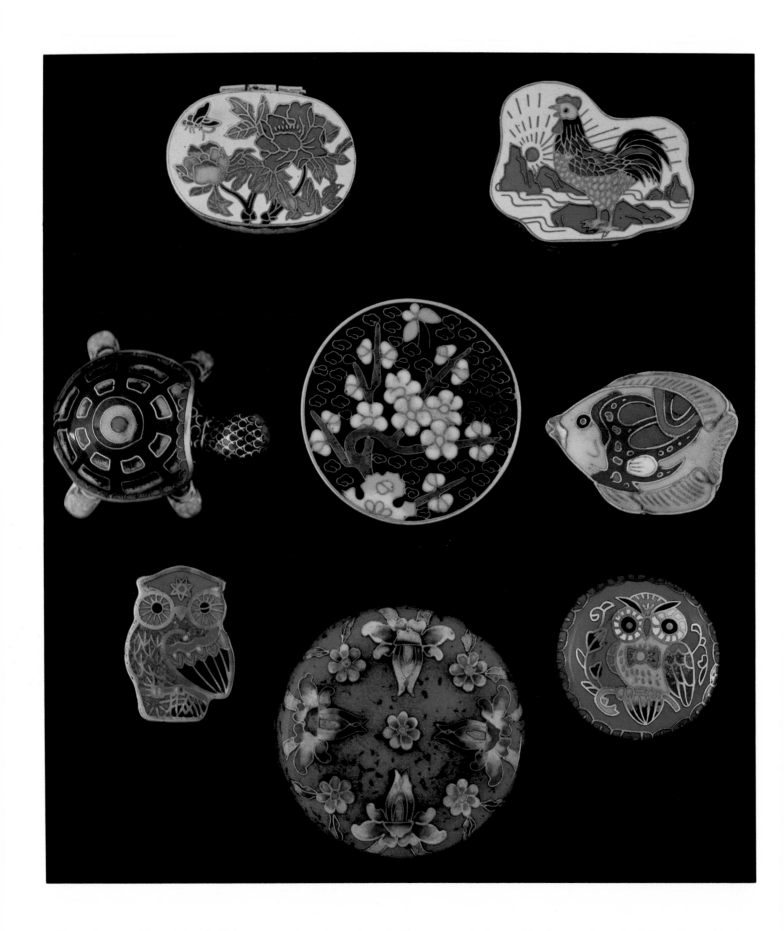

Enamel on a wide variety of pill boxes, small vases, perfume bottles, as pendants, napkin rings and candy dishes. Enamel looks bewitching on these miscellaneous objects.

Necklaces and Bracelets from China

Enamel adorns adornment. The application of enamel by Cloisonné technique to beads and round bracelets requires command of the more difficult technical processes.

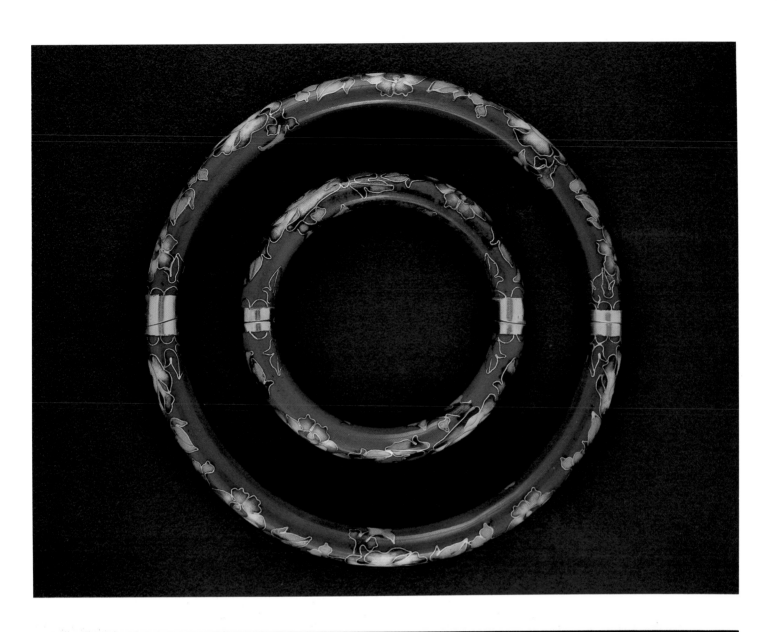

Enameled Everyday Objects

Above and opposite: ball-point pens, fountain pens, chopsticks, cigarette lighters etc. with innovative and effective Cloisonné enamel decoration (China).

The Five Steps in Making Window Enamel

1   2   3   4   5

1. Copper wires are soldered together over a plaster model and set in the right shape.
2. The plaster model is dissolved by acid; the copper wire form is silver-plated.
3. Enamel is placed in the spaces and fired. Filling and firing continues until all the openings are filled or closed.
4. Higher enamel is ground flush until the wires come into view again.
5. The whole piece is polished so that the enamel becomes glowing and transparent; then the wires are gold-plated.

Below: Tea bell, owl, and decorated Easter egg in window enamel, from China.
Upper right: Window enamel bowl in Viking ship form (Silver stamped "Deutschland"); the bowl below takes on an uncommonly effective glow from the different colors of its cells.

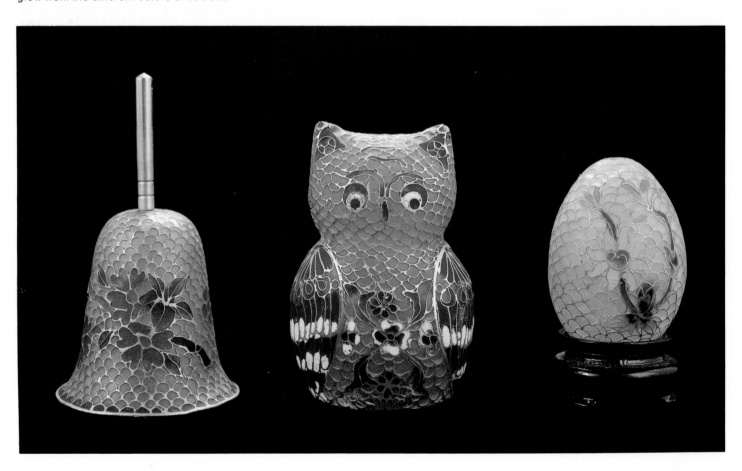

Enameled Thimbles

Thimbles are a special and very popular area of collecting. There are thimbles of a wide variety of materials, and since even ladies of the highest social levels did their own sewing in the salons of earlier days, it is not surprising that there are thimbles of valuable precious metals with costly decorations on the market.

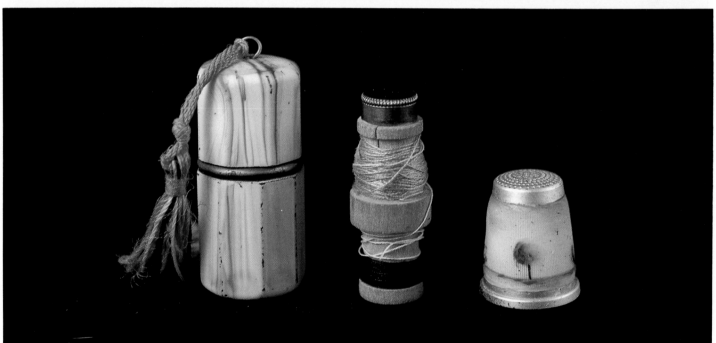

Above: Three German thimbles with translucent enamel over engraved silver. Underneath is a rarity: a colored sheet-metal case—surely intended for travel—with spool and thimble likewise of translucent enamel over silver.

Opposite page: There was obviously no limit to designs for thimbles. All these decorative thimbles in various animal shapes were made in China, using Cloisonné with strongly contrasting colors.

Cell-Enameled Plates

A plate made by the cell-enamel technique. The even colors that contrast strongly with each other are particularly well done. The smoothed and polished metal wires stand out clearly above the enamel layer.

Fayence Plate from Longwy

Typical enamel from Fayence, in late Art Nouveau, from the French city of Longwy. The flower ornamentation was applied so that the enamel remained pearly even after firing, which gave the plate a relieflike texture.

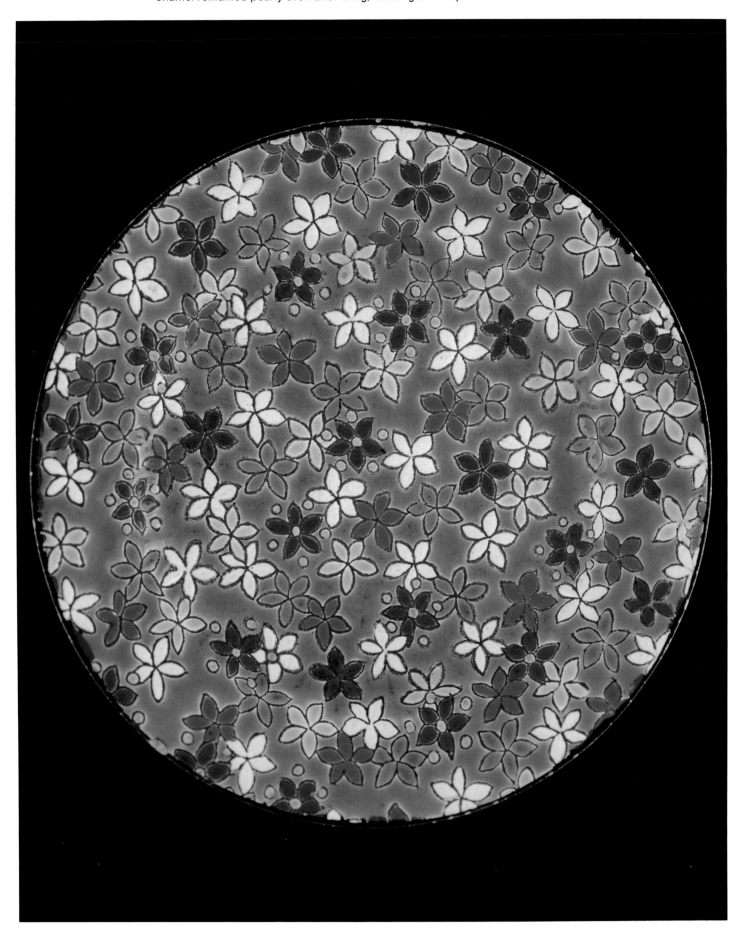

Enameled Photos and Photo Frames

Actual photographs, especially portraits, are often found on old drinking glasses as well as ceramic and porcelain beer mugs, or on oval porcelain discs with pictures of the dead, for use in cemeteries. Reproductions of photographs on wood or even leather are rarer; the latter type are called panotypes. All of these photographs on various materials involve a transfer process in which the photographic layer was transferred to the base material. The same process was used in photo enamel, with metal used as the base, but photo enamel of this kind is found very rarely because the reproduced picture could not have a screen, but involved only a pressing of the photographic material. A classic example of genuine photo-enamel is the picture on an oval copper plate below. A more accurate technical description is found in the glossary under the term "photo enamel." The beer mug is an example of photo enamel on glass.

Enameled photo frames are also a fascinating collecting area. Of the ones shown here, the small one at the left is of note, as it was made to take postage-stamp photos. The most beautiful enamel photo frames, though, surely came from the Fabergé studio and were favorite gifts of the nobility and wealthier bourgeoisie. The present-day collector can obtain genuine Fabergé photo frames as well as outstanding replicas.

Photo enamels on various surfaces; the family heirlooms of former days are desirable collectors' items today.

Pieces of glass enamel, made by the Millefiori technique.

Bottles and glasses with enamel color decor. The enamel used had to have a lower melting point than the glass itself. Such items were often made as status symbols and showpieces rather than for ordinary everyday use, which is why they are still in existence. Well-preserved complete sets are especially desirable among collectors.

Glossary

Absolon, William: This English craftsman worked as an independent enamel painter, gilder and glass dealer in Great Yarmouth, England between 1784 and 1815.

Adhesive: To guarantee a good grip on the metal for the enamel, bonding agents in the form of nickel and cobalt oxides must be applied. During firing, a chemical/physical reaction takes place between the metal surface and the enamel, causing a roughening of the metal and thus a better bonding of enamel and metal. German: Haftmittel.

Adhesive surface: Distilled water and *gum tragacanth* produce a slightly sticky solution which, when painted or sprayed on metal, improves the adhesion of the enamel powder before firing.

Ador, Jean-Pierre: This French goldsmith worked in St. Petersburg circa 1770-1785 and created, among others, notable enameled snuffboxes.

Aegean enamel: The culture of the Mediterranean area from 2600 to 150 B.C. is known as the Aegean Era. The oldest known enamelwork was found on Cyprus and was produced in this era. It consists of uncut cell enamel with raised wires.

Aging: As age increases, and also as a result of tempering at high temperatures, glass and enamel tend to change to their crystalline state under pressure. This process is also called *unglazing*. German: Alterung.

Ajour enamel: In this laborious technique, decors and patterns are created by cutting or filing out perforations in the sheet metal or by casting metal with open spaces. These openings are then coated with enamel powder and fired. The effect is similar to that of medieval church windows. See *Window enamel*.

Alpaca: Also called German silver, argentan, paktong. An alloy of copper (45-70%), zinc (8-45%) and nickel (8-28%). It is a good surface for enameling because it scarcely warps and has a high *melting point*.

Alumail: Specially mixed enamels for enameling aluminum. This technique has been known since 1950 and is chiefly used in industry and construction.

Annealing: Through heating, inner tensions in the enameled surface (metal), which occur through mechanical shaping, are eliminated. German: Ausglühen.

Application: The coating of the cells or pits of the metal surface with enamel. This is done with a brush or spatula, or by dipping, sieving, dusting or rinsing. German: Auftragen, Betragen, Eintragen.

Application technique: A metal plaque is dusted with enamel powder. Into this coating of powder pieces of enamel, in the form of lumps, ribbons or *Millefiori* are inserted. Metal particles and fireproof small bits can also be applied and melted in a single work process. German: Auflegetechnik.

Arcanist: (Latin: arcanum, meaning secret). The term used in the 18th Century for one who knew the secrets of porcelain and Fayence production. German: Arkanist.

Architectural enamel: Enameled steel and iron plates used as wall coverings add color and variety to the architectural scene. German: Architektur-Email.

Arlaud, Louis-Aimé: French miniature painter, born in 1752, who also painted with enamel colors.

Arte Vetraria: (Italian: the Art of Glass). Standard work on glass production by Antonio Neri, first published in Florence in 1612 and translated into German by J. Kunckel in 1679.

Base coloring: The metallic surface to be enameled is primed with a first layer of enamel, called the *base enamel*. German: Grundauftrag.

Base enamel: A special type of enamel including adhesive oxides, which provides for both good adhesion to the metal and equalization of tension between the metal and the covering enamel. German: Grundemail.

Basse-taille enameling: A pattern is engraved in metal and glazed with transparent enamel. This enameling technique was developed in Pisa in the 13th Century to decorate silver utensils.

Battersea: In 1753 S. T. Jannsen and his partners H. Delamain and *John Brooks* (see also) founded a workshop for enamel painting in the York House factory on the south bank of the Thames in London. Production of snuffboxes, containers, plaques and clocks began in the same year. Shepherd and generic scenes, portraits and landscapes on white backgrounds are typical of the early times. The work was based on Meissen painting. When the technique of *transfer printing* was adopted, gentle shadings of blue, sepia, red and purple dominated. The colors were painted thinly over the pressing and burned in, so that the first (pressed) pre-drawing shone through. Deep colors and a warm glow of enamel are characteristic of Battersea enamelwork. The firm's first engraver was S. F. Ravenet of Paris, who specialized in portrait boxes. Production continued in Battersea only to 1756; then the firm went bankrupt and its property was auctioned off.

Bechdolff, Johann Andreas: Born in Bautzen in 1734, he first worked as a porcelain and Fayence painter and later devoted himself completely to box painting on enamel. His enamelwork stands out because of its glowing colors and contrasts, which were otherwise not customary on enamel.

Beilby, William: He (1740-1819) and his sister Mary (1749-1797) were English enamel painters. They are best known for painting natural motifs on ale glasses.

Bienais, Martin Guillaume: A French goldsmith and enameler, ca. 1800-1832, who worked in Paris.

Bilston: Enamel production began in a small factory in Bilston, Staffordshire, circa 1750. Generic scenes, shepherd idylls, flowers, fruits, insects, landscapes and even architecture were the favorite motifs. The work was based particularly on that of *Sèvres. It is possible to classify the products chronologically on the basis of the colors used:*
1750-1757: dark blue
1757-1760: pea green
1760-1770: turquoise and wine red
1770-1780: silver, yellow and red gold
After 1780 only mechanically mass-produced articles were made.

Birch, William Russell: English enamel painter who worked between 1775 and 1794.

Birmingham: Production of painted enamel boxes began in this English city about 1750. Romantic French landscapes and groups are characteristic of the early work. After 1760 transfer printing was used in painting the articles. The patterns were applied with used plates from Battersea, which were used often, resulting in weaker designs. When compared to Battersea enamel, the articles were colored with thick layers of enamel. Birmingham enamelwork was very much in demand, and much of it was exported.

Black solder: An enamel color consisting of a pulverized mixture of iron and copper with glass powder. German: Schwarzlot.

Black solder painting: Painting on glass or ceramics, sometimes also on porcelain, with black solder, sometimes in combination with iron-red painting and gilding. The chief exponent of this technique in Germany was the painter J. Schaper, who was much imitated in the 17th and 18th Centuries.

Blois enamel: J. Touth developed a special technique of enamel painting in Blois in 1630. Flowers, fruits and scenes were painted as miniatures. Later the center of this painting moved to Geneva.

Blood enamel: In Celtic artistic work of the newer iron age (450-50 B.C.), coral was first used, and later opaque red enamel—called "blood enamel." The work was prepared like *groove enameling* and then filled with red enamel and fired. German: Blut-Email.

Boit, Charles: 1662-1727, born in Stockholm, he was England's first significant enamel painter. His specialty was portraits.

Bone, Henry: 1755-1834, English enamel painter, the most successful of his time.

Borax: Borax supplies the enamel with the B2O3 boron trioxide needed for its composition. This oxide is a good glass former. It also facilitates the dissolution of the metallic oxides added for coloring and lowers the enamel's melting point.

Bott, Thomas John: 1829-1870. He worked in Worcester, England, primarily painting white decorations on a black background. Thus he was able to imitate the effects of 16th-Century painter's enamel from *Limoges.*

Brass and bronze: Enamel adheres to these metals poorly. Thus before they are enameled, a roughened surface must be created. But even then, results equal to those achieved with copper cannot be attained.

Bristol glass: In the 18th Century Bristol was a significant English glass center. Its opaque white milk glass, resembling porcelain, was famous. It was also decorated with enamel colors. Blue glass also was produced and often painted with white enamel that could be fired onto the glass excellently because of its high surface tension.

Bronze: Collective term for a group of thick, hard alloys of copper (over 60%), tin (20-30%) and other metals. Bronzes have good elasticity and working properties, and high resistance to damage and corrosion. They are formed by casting. The Celts (3rd Century B.C.) were already making and enameling weapons and jewelry of bronze.

Brooks, John: Irish designer and copperplate artist, active from 1710 to after 1756, who developed the process of duplicating drawings on porcelain and enamel. In 1753 he was one of the founders of the enamel factory in *Battersea.*

Brown varnish: See *Email brun.*

Burning through: A surface to be enameled is covered with stiff enamel with a hot melting point. Afterward softer enamel with a lower melting point, in a contrasting color, is applied. The work is fired briefly and at very high temperatures. Thus the softer type of enamel sinks partly into the stiffer underlying layer; under some conditions the enamel layer is penetrated to the metal. In the process, interesting color mixtures and charmingly colored structures not attainable by other techniques are formed. German: Durchbrenntechnik.

Byzantine enamel: Byzantine enamel forms the basis for the development of enamel art in medieval Europe. Enamel first played a central role in goldsmithing. Cell enameling in particular was developed to its perfection.

Camaïeu: Cameo-style painting on various backgrounds, also on glass, porcelain, and enamel, in several shades of a single color. A special form is gray-on-gray painting (see Grisaille).

Cambrian pottery: Household wares firm founded in Swansea in 1767. Cream-colored stoneware and pearlware were made and printed in black, brown or blue. As of 1802, pearlware was also decorated with animals, fruits and flowers in enamel painting. In the mid-19th Century, terracotta vases and figures with black enamel painting were also made.

Canton enamel: During the reign of the Emperor K'ang Hsi, missionaries brought works and painter's enamel to China and had the pieces copied. The technique involved was also used there to decorate ceramics.

Casali, Antonio: An Italian Majolica painter from Lodi, active from the middle to the end of the 18th Century. He decorated tin-glazed articles with enamel painting, imitating Chinese porcelain.

Casting: Metals are melted and cast in prepared forms. The gold figures for plastic enameling are usually produced by the lost wax process *(à cire perdu).* This is the oldest form of casting with a lost form, which allows not only the casting of small parts, but also seamless casting of large-surface plastic pieces. To produce small plastic objects and pieces of jewelry, models are made of wax, placed in casting sand, and the wax is then melted out. The metal is then cast in the resulting empty space, and the casting sand is chipped off after cooling. The cleaned metal piece can then be further improved, among other ways, by enameling; see also *plastic enameling.* For plaques and flat plates it is possible to use a form of wood, copper, plaster etc. over and over as a model. This model is pressed into casting sand and the resulting relief is cast in metal. In this way, many identical surfaces for pit, bridge, or Cloisonné work can be produced. The laborious work of bending and soldering the wires or etching and cutting out the pits is thus eliminated. German: Guss.

Cell glaze: see *Cloisonné.*

Celtic enamel: In Celtic handicrafts, the enameling art reached a peak in the newer Stone Age (450 B.C. to the beginning of the Christian era). Its greatest development occurred in England in the First Century A.D.

Cement: see *gum arabic, gum tragacanth.* German: Bindemittel.

Ceramics: Collective term for products made of fired clay. They can be divided into products with porous textures (pottery, terra-cotta, Fayence, earthenware) and stoneware and the various types of porcelain with thick, sintered textures. German: Keramik.

Chaffers, Richard: English potter working in Liverpool, 1731-1765. He produced good porcelain, which he usually printed with blue decors, but also with polychromatic Chinoiserie in enamel painting.

Champlevé: See also *pit enameling;* This technique reached its first peak in the Celtic art of the First to Third Century. In the high Middle Ages it was used by significant artists such as Nikolaus of Verdun and the famous schools in Namur, Paris, Limoges, Milan, Maas and Reims.

Chasing: The metal piece is laid on a soft material (sandbag, potting soil), and various lines, patterns etc. are beaten into it with the chisel or punching hammer. In making figures, details such as hair, beards, fur and wrinkles are chased.

Chinese enameling: The Cloisonné technique reached China from the West in the 14th Century, under the Mongol Yuan Dynasty. In the following Ming era (1368-1644) it attained higher accuracy and refinement. During the Ching-t'ai period (1450-1456) blue was especially popular. For that reason Cloisonné is sometimes called "Ching t'ai blue" even today. The Chinese used copper as the metal surface. The wires were also made of this metal; after firing, they were gilded. On the other hand, gold was preferred in Byzantium. The artistry of this enamel technique is still preserved and practiced in China today, and marketed in an incredible variety of forms and colors.

Claire, Godefroid de: The most famous goldsmith and enameler of his time. He worked in Huy in the 12th Century, creating, among others, the Alexander Reliquary of Brussels, the Heribert Shrine in Deutz and the Cross of St. Omer.

Claire, Martin Guillaume: French goldsmith and enameler, working in Paris ca. 1800-1832.

Clauce, Isaak Jakob: 1728-1803, German miniature painter, who painted for Meissen Porcelain, among others. Enamel paintings by him are also known.

Clobbering: English term for the additional painting of blue and white porcelain with enamel colors. In London between 1824 and 1850, porcelain from Chelsea and Worcester was decorated in this way. In Meissen too, in the mid-18th Century, porcelain by Ferner was decorated with enamel colors.

Cloisonné: This is probably the oldest known enameling technique. Small metal wires are soldered or glued to a metal surface and form the outlines of the decor. The resulting cells are filled with enamel, fired and smoothed. This technique, already known in Egypt, Greece, Rome, and the Orient, reached its peak in the early Middle Ages in Byzantium and in Germany in the early days of the Holy Roman Empire. Precious medieval works include the Pala d'Oro of San Marco in Venice and the Staurothek in the cathedral of Limburg on the Lahn. German: Zellenschmelz, Senk-schmelz, Vollschmelz.

Cloisonné technique: Probably the earliest and most valuable of all enameling techniques. It can be done, as desired, with either opaque or transparent enamel. Any metal surface used in Cloisonne work is surrounded by a protective edge, called the "Bördel." The fillet wires of which the figures and cells are formed are made by flattening or rolling copper wire. Before the wires are attached, the surface should be covered with counter-enamel, which is also possible with a later firing. Now the wires are bent to fit the pattern, using forceps and small pliers. The individual parts must then be placed on the surface like pieces of a puzzle. The wires should be bent in a way that will prevent their falling off. A thin liquid adhesive such as gum tragacanth helps to hold them. The individual parts are dipped in the adhesive and then arranged on the surface. As long as the adhesive is wet, the outline of the pattern can always be corrected. It is safer to sift fondant onto the surface and then to put the wires on. In firing, the wires are melted and thus fixed in place. For uneven or vertical surfaces (bowls, vases, boxes) the wires must be hard-soldered to the surface in places. Along the edges, cast-in ridges in the surface can allow large-quantity glazing of the same pattern. The fillets are already cast into the form and must not be laboriously bent and soldered on. But these mass-produced goods lack the light, delicate quality and look rather dull. The previously prepared copper surface can also be silvered galvanically before being enameled. Thus no disturbing layer of oxide forms during firing. In the case of copper, this must be removed before each new coating with enamel. Now the small cells and larger areas are coated with a spatula or a fine-pointed brush. Small portions of liquid enamel are taken up with the spatula or brush and the cells are filled with them. When the cells are brim-full of the desired colors, then the work is allowed to dry and then fired. Thus the enamel sinks into the cells. How often the work has to be applied and melted depends on the height of the cell bridges, but also on the amount of enamel that can be applied after every process. Repeated thin application lets the colors melt through better, which enhances their effect. When all cells have been filled with enamel after the last firing, the surface is smoothed down evenly with a carborundum stone under spraying, cooling water. This must be done very delicately. The enamel shimmers uniformly under water: all cell wires lie free now. But after drying it shows that the abrasive stone has left scratches, and small pits or holes may also be found. That means the whole surface has to be cleaned in thin spirits of salmiac, rinsed, dried, and have new color particles applied and be fired again. Now work proceeds with finer and finer material, and finally the beauty of the matte finish is brought out by polishing. The finished product now can be galvanically silvered or gilded with no problems.

Coating: Cells not filled with colored enamel are coated with fondant to attain an evenly thick covering of enamel. Transparent enamels can also be coated with fondant so that they do not become too dark as a result of frequent application (thickness of layers). German: Uberfangen.

Cobalt blue: see *Derby blue.*

Cold enameling: Cold painting with a synthetic paint, which gives a similar appearance to enamelwork after drying. This is a *painting* process, not an *enameling* process. German: Kalt-email.

Cold painting: Cold painting on glass came back into fashion in the Biedermeier era. Enamel colors (molten glass colored by metallic oxides) are pulverized and stirred into a semiliquid mass with an adhesive (varnish, oil). This mass is then painted onto the glass but not fired. This type of painting is very sensitive to use; it flakes off easily. Ceramics and enameled objects were also decorated with cold painting. German: Kaltmalerei.

Color test: An enamel is melted on various metals to document its differences in color effect and translucence. German: Farbtest.

Copper: Metal with a reddish metallic luster, chemical symbol *Cu,* from Latin *cuprum.* It is flexible, durable and can be embossed easily, taking on a variety of different forms.

Copper and its alloys (with zinc + brass, with tin + bronze) are, along with silver and gold, the most important metals used in decorative enameling. German: Kupfer.

Corplet, Charles Alfred: 1827-1894; French enamel painter and restorer.

Coteau, Jean: French enamel painter, born 1739, no trace remains after 1812; decorated snuffboxes in Louis-seize style and developed a method of translucent enamel painting with the use of paillons (see *paillon).*

Cotes, Samuel: 1734-1818; English enamel painter who specialized in portraits and miniatures.

Counter-enamel: To equalize tension, the back of the enameled surface likewise has enamel applied to it. German: Gegenemail, conteremail; French: Contre-email.

Covering lacquer: German: Abdecklack. See *Etched surface.*

Craquele enamel: A special, fine-ground opaque enamel that is applied to a pre-fired enamel surface in a thick liquid solution and is then fired at rather high temperatures (ca. 870 degrees Celcius). When fired, this upper layer breaks and leaves a fine network of cracks.

Crosse, Richard: 1742-1810; English enamel painter, who painted chiefly miniatures and portraits.

Cuerda seca: Islamic potters developed this technique to prevent glazes and molten enamels from running into each other. The outlines of the decor are applied with manganese purple mixed with animal fats and the spaces are filled with enamel. In firing, the animal fats burn up but act as a barrier, preventing the colors from mixing. This technique has been known in North Africa since the 11th Century, in Persia and Valencia since the 14th Century, and in Spain since about 1500. See also *Line technique.*

Damacene intarsia: Metal-in-metal intarsia; fine grooves are cut or etched in objects of iron, bronze or brass to form the pattern, and wires of gold, silver or copper are set into them and hammered or polished even with the surface. This technique arose in the Near East around Damascus, hence its name. German: Damaszieren, Tauchieren.

Darkening agents: The usual darkening agents are the oxides of titanium, zircon, antimony and molybdenum, as well as antimony compounds (SnO_2, ZnO, CeO, TiO_2, Sb_2S_3 and As_2S_3) and ceramic coloring materials (inorganiz oxides) such as fluorite and cryolite. In a narrower sense, the darkening agents darken white to the greatest possible degree. Bone-ash (calcium phosphate), formerly used for darkening, has lost its significance today. German: Trübungsmittel.

Davenport, John: English potter from Staffordshire, born 1766. Among his creations are Toby jugs of cream-colored earthenware, painted in glowing enamel colors.

Decorative enamel: Enamel with a low melting point and many melting agents, mostly containing lead, and with little quartz content compared to technical enamel. German: Schmuck-Email.

Deep-cut enameling: German: Tiefschnitt-schmelz. See *Relief enameling.*

Degreasing: So that the enamel adheres well to the metal surface, the latter must be grease-free. This can be done mechanically by scouring, annealing and subsequent pickling, with ultrasound, electrolytically or chemically by leaching or using solvents. German: Entfetten.

Derby blue: A blue enamel color based on royal blue. Also royal blue, cobalt blue.

Dinglinger, Johann Melchior: 1665-1731; The most famous jeweler of his time, he worked as the leader of a group of court jewelers for August the Strong. Significant enameled jewelry ranks among his works.

Discoloration: Glass can be discolored by oxidizing agents such as PbO_2, KNO_3, As_2O_3, MnO_2, or by reduction through carbon substances. Discoloration was already known in earlier times and later forgotten. In the 16th Century it was rediscovered by the Venetians. German: Entfärben.

Ducrollay, Jean: A French goldsmith, 1709 to after 1761, who produced, among other objects, especially lovely boxes with enamel decor.

Duesbury, William: An English ceramic painter, 1725-1786, who also decorated salt-glazed goods with enamel colors.

Dusting technique: Fine-ground colors are thickened with water and gum arabic to make a sprayable paste. A stencil is cut for each individual color, covering those surfaces to which that color should not be applied. The stencil is put in place and the color sprayed over it. After drying, the next stencil and color are applied. Overlays and mixed colors are also possible. Special effects can also be achieved by regulating the spray. After drying, all colored layers are fired at once. German: Spritz-Technik.

Earthenware: Light ceramics that give a light effect as does porcelain but is considerably thicker and more likely to break. The material is porous and becomes watertight only when glazed.

Edkins, Michael: English enamel painter, 1734-1811; at first he painted decorative objects made of English Delftware; as of 1762 he specialized in enamel painting of opaque white and blue glass.

Egyptian enamel: Science has proved that the Egyptians, at least at the time of Tutankhamen, knew enameling in the *strict sense.* Many pieces of jewelry and masks from pharaohnic times were made by fitting and gluing prepared pieces of glass into gold cloisons or glass paste was pressed in and then hardened (see *Inlay work*). They look like enamel but, since the glass was not melted *in situ,* they do not fit our definition of enamel. German: Agyptisches Email.

Eilbertus: A monk at St. Pantaleon Abbey in Cologne; he lived in the 12th Century and worked as a goldsmith and enameler.

Electroplating: G. R. Elkington patented the process in 1840. By electrolysis, a layer of metal is applied to another metal to make it more resistant to corrosion or give its surface a more beautiful and valuable appearance. German: Elektroplattieren.

Email à jour: Window enamel. See *Email de plique à jour.*

Email brun: see *Varnish firing.*

Email champlevé: See *Champlevé.*

Email de basse taille: See *Email en basse taille* and *Silver relief.*

Email des peintres: In this technique the metal serves only as a surface for painting. One covers it first with an opaque monotone layer of enamel and then paints the picture on it by applying and firing subsequent enamel layers. Often parts of the metal surface are left free. See *Painter's enamel.*

Email de plique à jour: Perforations of the metal glazed with transparent enamel so light can shine through unhampered. Also a special type of Cloisonné in which the wires give the transparent enamel form and outline without a metal background. This unusual technique was used in China since the 11th Century and in Russia and Scandinavia. See also *Transparent Cloisonné, Window enamel.*

Email en basse taille: Translucent *silver enamel,* especially popular in 14th- and 15th-Century Gothic art. The design is carved or cut into the silver surface and glazed with transparent enamel, so the silver background reflects light.

Email en resille sur verre: *Glass enamel.* A pattern is engraved in blue glass or colored crystal glass and applied in gold leaf. Then the cuts are filled with transparent colored enamel and fired. Naturally the melting point of the enamel must be lower than that of glass. This technique was used in France and Germany, probably in the early 17th Century, for jewelry and miniatures, but was soon given up since the laborious enamel melting resulted in much breaking and cracking.

Email en ronde bosse: See *plastic enameling.*

Email en taille d'épargne: see *gold pit enameling.*

Email sur bisquit: In Chinese *San ts'ai* ware the term for Chinese three-color ceramics used since the 16th Century. The material is brownish-yellow or gray stoneware, or sometimes porcelain. The colors are applied in fields bordered by raised clay lines (similar to fillet enamel) or set directly beside each other. By contour and interior drawing (usually in manganese brown), nuances are worked out.

Email translucide de basse taille: See *Basse-taille.*

Email veloute: This technique was developed by Demond Lachenal. To achieve gloss/matte effects, all the portions of an enameled plate that are to remain gloss are covered with shellac. After drying, the piece is bathed in a weak liquid acid solution.

After the shellac is removed, interesting contrasts between matte and gloss are seen.

Embossing: Gold, silver, and copper can be shaped cold by embossing without losing their elasticity and resistance. With short hammer blows, the sheet metal is hammered on one or both sides, while constantly being turned and pressed, until the desired form is attained. Along with free embossing, embossing of fixed forms is also possible, but usually only for flat reliefs.

En plein enameling: The enamel is applied to the entire surface of the object.

Enamel: The term comes from the Latin "smaltum": to melt, German "schmelzen," Old High German "smelzan," Italian "smalto" and French "email." The French word has also been adopted in German. Glassy, chemically somewhat resistant but somewhat shock-sensitive thin overlays on utensils of sheet or cast iron (kitchen utensils as well as industrial apparatus), or as a decoration on copper, tombac, brass, gold, silver, glass, or clay articles. By scientific definition, enamel is a solidifying glassy mass of inorganic substances, mainly oxides, which comes into existence through melting or fritting and is applied in one or more layers on **metal.** More simply stated: **Enamel is a layer of glass melted onto metal.** The end products are similar to those made of glass. Technically, enamel is ground colored glass that is melted onto a metal surface in an oven. The ground glass, which is applied to the surface in powdered form, is dull, with none of the fire that it shows only after firing.

Enamel advertising signs Enameled signs are a special form of advertising placards. In 1859 the Englishman Benjamin Baugh opened the first known enamel sign factory in Birmingham. When the first such factory in Germany was opened is not definitely known, but the firm of Robert Dold in Ortenberg, Baden was first mentioned in 1894. Like the firm of Boos & Hahn, which was founded in 1917, also in Ortenberg, it ranks among the most important German manufacturers. The first enameled signs were made of cast iron. This coarse material is given a smooth surface by sandblasting and is then dipped into the liquid enamel and dried. The sign is fired in an oven heated to 900 degrees Celcius and then, while still hot, the design is applied in enamel powder by means of a stencil, melting immediately and forming an even

layer of enamel. For every additional color the sign must again be heated to 900 degrees Celcius and have color applied with a stencil. The relieflike surface is easy to define with the enamel layers that lie one over another. Toward 1920 sheet steel came on the market, which allowed the enameling process to be simplified. The thickness of enamel gives a hint of the age. Older signs have an enamel layer several millimeters thicker, for at that time the intensity of the color could be attained only through thickness and not with metal oxides.

Enamel boxes: The great era of enameled boxes began in the first half of the 18th Century. Its development paralleled that of the porcelain box, since they were often made just to imitate the porcelain that was very costly and desirable then. With their white or pastel-colored backgrounds and bright-colored painting, they have a resemblance to porcelain. Several outstanding porcelain or Fayence painters were also enamel painters, for example C. C. Hunger, E. F. Höroldt and A. Bechdolff in Germany and J. Brecheisen in Denmark. The most important German centers for the production of enameled boxes were Augsburg, Berlin, and Dresden. Those in England were Battersea in London, Birmingham, and Bilston and Wednesbury in Staffordshire. While early French enameled boxes were almost always gold, in England and other countries the less expensive copper was used from the start.

Enamel colors: Melting glass colored by metal oxides are used to color porcelain, earthenware, stoneware, Fayence, glass and enamel. They are applied to the glaze (and thus also called *overglazing colors)* and fired in *muffle ovens* (and thus called *muffle colors)* at 500 to 1000 degrees Celcius, depending on the nature of the object.

Enamel frit: A frit comes about through an unfinished melting process using simple raw materials in powdered form. In practice, lumps of raw enamel are also called frit.

Enamel glass: Hollow glass in various forms is painted with enamel colors and then fired. This type of decoration was already known and loved in Imperial Rome, Byzantium, in the Middle Ages (Venetian kilns) and in Germany in the 16th to 18th and 19th Centuries (German kilns). German: Emailglas.

Enamel granulation: Small silver balls are

melted onto enamel.

Enamel painting: In its broadest sense, this includes all processes of artistic application and subsequent melting of enamel and metal oxide colors on porcelain, ceramics, enamel or glass. German: Emailmalerei.

Enamel painting on ceramics: This technique was first used on earthenware of the later Sung era in northern China. In the 14th and 15th Centuries it was already found on Persian earthenware. In Europe this decorative technique was first used on stoneware in Kreussen, later also on porcelain and Fayence, especially in Strassburg. In China it developed into five-color painting on porcelain (wu ts'ai) by the Ming era.

Enamel painting on enamel: Since every enamel color requires its own firing temperature, a multicolored enamel painting must be fired very often. To shorten this long procedure, the surface to be enameled was covered with a solid bright layer of enamel, on which enamel colors (pure metal oxide colors) could be painted and then melted. The fine-ground enamel colors used are mixed with thick oil on a glass plate and then painted on with a brush. Enamel painting with only oxide colors is usually matte and must still be covered with colorless enamel. This technique is similar to porcelain painting and, as there, involves miniature painting and especially fine work. It was used particularly on boxes, clocks, etc., in the 18th Century.

Enamel painting on glass: From Egypt it came to Syria and Gaul around the 12th to 14th Centuries, immortalized in splendid Islamic glasses. In Renaissance Venice it reached a peak (the Berovieri family), then peaked again in Bohemia in the Baroque era (Preiser, Schapper). In the 16th Century, richly decorated Imperial Eagle tankards were made there, as were ox-head glasses in the Fichtelgebirge area and Halloren tankards in Thuringia. Significant products of this era are the Bohemian four-sided bottles with tin screw tops. From the early 18th to the mid-19th Century, folk-type enamel painting on glass was done most often by artisans and farmers. The bottles and tankards could be ordered with motifs and sayings of love, marriage, hunting and handicrafts. The background colors were glowing white, the dark green of woodland glass, dark moss green, cobalt blue and rarely also manganese violet. At the end of the 18th Century milk-glass was also used.

In 19th-Century Vienna and Dresden, transparent enamel painting was done on glass (Mohn, Kothgasser); in late 17th- and 18th-Century England a high point was reached by Beilby, Edkins and Giles.

Enamel powder: Enamel is available not only in powdered form but also in lumps, granules, balls, ribbons and as colorful round panels, called *Millefiori* or thousand-flowers.

Enamel removal: Removing the enamel layer from the background. This is done either mechanically by hammering or by heating and then chilling, or by the use of acidic (sulfuric or fluoritic) or alkaline (natrium hydroxide) chemicals. Sandblasting and ultrasound are also being used more and more often. German: Entemaillieren.

Enamel tombac: A copper-zinc alloy with 5-6% zinc. German: Emailtombak.

Enamel, types of: Gloss- and matte-firing enamels must be differentiated. Gloss-firing enamels can be opaque, opalescent or transparent, while matte-firing enamels are only opaque.

Enameler: A skilled industrial trade, or a part of the goldsmith's training. German: Emaillierer, French: Emailleur.

Enameling: Basically this term means nothing more than glazing specially prepared glass onto metal by means of heat. It should be noted that enamel conducts heat better than ordinary glass, so that no tensions arise between it and the metal during cooling. The melting point must also be lower, for it must melt sooner than the metal surface. Enameling can be done with either wet or powdered enamel, and is usually done in several layers. German: Emaillieren.

Enameling mistakes: They can occur through faults in the metal surface, by under- or overfiring of enamel, and through contamination, and take such forms as fishscales, cracks, pimples, sulfide depressions and the like. Possible mistakes:

Metal parts melt	Oven too hot
White has green spots	Contaminated with copper
Colors have black spots	Iron tinder spattered
Gray spots	Bits left from cutting
Enamel or unsuitable metal	porous
Colorless enamel	Too finely ground
Becomes milky	Washed too little, applied too thickly, fired
Enamel has cracks	Enamel spread unevenly under heat, tension between enamel and metal, wrong metal thickness, not enough counter-enamel
Dull gray enamel	Not rinsed enough
Plates very warped	Wrong application build-up, not enough counter enamel.
Enamel is bubbly	Dirt particles burned, different enamels inter-reacted.

Enameling surface: Plaque of raw material that bears the enamel, usually of copper, tombac, gold or silver, forming the background on which the enamel powder is fired. Other metals can also be used (for industrial uses also iron, steel and aluminum) if their melting points are over 900 degrees C. German: Emailträger.

Engraving: Probably the oldest technique of decorating metal, it was already widespread in the 13th and 14th Centuries and later in the 17th and 18th Centuries. Unlike chasing, where the lines are struck in, the design is engraved on the smooth object with a sharp tool, as in copperplate making. To make the engraved design more visible, black coloring is rubbed into the cuts; see *Niello*. German: Gravieren.

Essex, William: English enamel painter, 1784-1869, who painted landscapes, classical motifs and historical portraits for Queen Victoria.

Etched surface; covering lacquer: Prepared covering lacquers contain, among other things, wax, pitch, colophonium, mastic,

Dammar resin, turpentine. The copper areas not to be etched can also be covered only with wax. German: Atzgrund.

Etching: The metal is partly covered with, for example, asphalt lacquer, and then dipped in an etching solution. The uncovered areas are thus etched off and a low relief results. When left in the etching solution longer, deeper pits can be etched out of the metal; they are then filled with enamel and fired, producing *Champlevè.* German: Atzen.

Etching copper and tombac: To be able to etch a design or a pit deep in a metal, the surface of these lines has to be coated with a covering layer to protect it from the etching chemicals. A lacquer is generally applied to the surface with a brush as a covering agent. After etching, the painted-on motif stands out. If the whole plate is coated with covering lacquer and only the design is scratched out, then only these lines will be etched, and the surfaces remain. This gives a delicate appearance. There is also the possibility of using a synthetic wax like Ceresine as a cover. The entire surface can be covered with a layer of wax and the design can then be scratched out, or the design can be applied to the warmed surface with a tainting as in batik. A third possibility consists of applying wax ribbons (cotton fibers dipped in molten wax). Each of these methods leaves its own type of design. Objects thus prepared can then be etched. The object can be placed in sulfuric acid or iron (+ 3) chloride solution and left for 40 to 60 minutes. When the etching is deep enough, the object is washed thoroughly in water and the wax melted off.

Etching fluids: To etch copper and its alloys, acid mixtures are used, such as thinned sulfuric acid or iron (+ 3) chloride solution, to which hydrochloric acid or potassium chlorate can be added. German: Atzmittel.

Fabergè, Carl: Russian-French jeweler and goldsmith, 1846-1920, who prepared particularly elaborate enameled jewels for the Tsars. His father had founded a jewelry business in St. Petersburg in 1842, which he and his four sons developed into an internationally famous firm with branches in Moscow, Odessa, Kiev, London and Paris.

Feldspar: Alkaline or aluminum silicate, important component of porcelain, glazes and enamel. It decreases drying and evap-oration and aids the flow during firing. German: Feldspat.

Ferner, F. J.: The Bohemian court painter painted flawed porcelain from Meissen with enamel colors in the mid-18th Century.

Filigree: Latin *filum:* wire; filigree is a decor of thin gold, silver, gilded or silvered wires that can be smoothed. They are formed into spirals or tendrils and plaited to form grilles, then soldered to the object to be decorated. In later times the wires have been worked into artistically intertwined plaits without a basis and only the intersections are soldered together or soldered like a net over excavations in metal *(ajour).* Self-supporting pierced decorations are formed. This technique was already used in antiquity, especially by the Etruscans, and was very popular in the Middle Ages and the Baroque era. German: Filigran.

Filigree enamel: Goldsmithing work of gold, silver or copper wire combined and melted with enamel, a technique often used in Hungary. German:Filigran-Email.

Fillet enamel: A form of cell enameling in which the enamel is not raised to the full height of the wires. Since it rises to some extent along the sides of the slightly curved wires, small reflective curves are created. German: Steg-email.

Finishing: The final preparation of a piece of work after the last firing, for example, grinding, polishing, gilding.

Finishing firing: *Gloss firing* after cutting, at ca. 900 degrees Celcius, until the surface has become glossy. German: Schlussbrand.

Firescale: A serious mistake in enameling: The molten enamel breaks off the metal surface in flakes. German: Fischschuppen.

Firing: After applying the enamel and letting it dry on the metal, the object is fired smooth in a *muffle oven.* The firing process creates a close bonding of enamel and metal. German: Brennen.

Firing forms: The success of a work often depends on these firing aids. When enamel is put into an oven, it can only be on a surface that allows it to be moved into and out of the glowing cauldron. These firing forms must fit the object, as they should only touch small bits of its surface. They are needed in varying shapes and sizes. In the heat of the firing process, bits of iron oxide spring off forms of ordinary steel; they can easily burn into the enamel and remain as black spots. It is best to use forms of stainless steel, since they do not react thus and so leave no black spots. German: Brennroste.

Firing temperatures: For enamel, they are normally between 800 and 950 degrees Celcius (see also *Melting point.)*

Firing time: The fired object is observed and taken out of the oven as soon as the enamel has melted smooth and glossy. The length of firing depends very much on the temperature in the muffle oven, the thickness of the metal surface, the melting temperature of the enamel and the thickness of the applied enamel. In general, it can be said that at a temperature of 850 degrees Celcius and a material thickness of 1 to 1.5 mm, firing must last 3 to 4 minutes.

Flint enamelware: American pottery glazed by a process patented by Fenton in 1849. Pulverized copper or cobalt oxide is applied to Rockingham glaze before burning and only then fired. The results are glowing colors of green, orange-yellow and blue on a brown background.

Flux: To attain thin, flowing melting, various fluxes are added to the raw materials of enamel. They include borax, soda, saltpeter and potash. Lead, barium and zinc compounds are also added to decorative enamels, but they are banned from cookware for health reasons. German: Fluss-mittel.

Foils: To increase the light transmission of *translucent enamels,* foils of gold or silver are used as backgrounds. These sheets of metal are 1 to 2 microns thick (1 micron + 1/1000 mm).

Fondant: A colorless *transparent enamel* used to underlie colored layers, even out high and deep spots in individual color areas or wires, or to protect and improve entire enameled objects. There are two basic types: a harder fondant to underlie colored areas—it melts only at high temperatures of about 900 degrees Celsius; and a softer fondant to cover individual areas or the whole piece—it melts somewhat lower than most colors.

Foucher: A French enamel painter, born 1662, who worked in Blois, painting mainly mythological scenes on clock casings.

Frit: See *Enamel frit.*

Fromery: Before 1750 the Berlin workshops were the main producers of enamelware in Europe. At Fromery the technique of gold relief application on enameled boxes, including with translucent enamel, was mastered superbly. After 1770 production declined, as jewelry, porcelain, and English enamelware become more popular.

Full glaze: The *cell glaze* covers the entire surface of the object to its outer rim, as opposed to *sink enameling.*

Galvanizing: An anode of precious metal and the base-metal object to be galvanized are immersed in an electrolytic bath and the current is turned on. The ions travel from the anode to the base metal, which functions as a cathode, and coat it. German: Galvanisieren.

Geneva enamel: A technique of enamel painting developed in Blois, often used to enamel clocks. Its center moved to Geneva around 1700.

German silver: Argentan, Alpaca, an alloy of nickel, copper and zinc, also used as a surface for enamel. German: Neusilber.

Gilding: German: Vergoldung. Types:

a. *Cold gilding* or water gilding: gold leaf is applied with an adhesive; it wears off easily.

b. *Lacquer gilding:* Pulverized gold leaf is mixed with lacquer and applied; it wears off easily.

c. *Oil gilding:* Pulverized gold leaf is mixed with linseed oil, gum arabic and mastic and applied. The gilding is somewhat more durable but matte.

d. *Honey gilding:* Pulverized gold leaf is mixed with honey and painted on glass or ceramics and fired at a low temperature. The resulting surface is hard enough for further working.

e. *Amber gilding:* Gold leaf is applied with a lacquer made of pulverized amber and fired at low temperatures.

f. *Molten gold:* Gold is dissolved in aqua regia, precipitated as a brown powder in iron vitriol, and pulverized with porcelain liquefier. The resulting gold paint is painted on as a glazing color and fired. The gilding takes on a gold ochre tone when fired.

g. *Gloss gold:* The gold application includes some 10-15% gold in oil. It is applied with a brush and fired. After firing it comes out of the oven glossy.

h. *Polished gold:* It exists in two forms, as powdered gold in bismuth oxide as a flux and as a 20% liquid. The strong liquid must be warmed and shaken well before applying. After firing, it comes out of the oven matte and lime brown. Now it is polished with agate polishers, bundled glass fibers or semiliquid chalk. The gold coating corresponds completely to its undercoat and thus is glossy on a gloss undercoat and matte on a matte one.

i. *Fire gilding:* Gold powder is mixed with mercury to make an amalgam and applied. On subsequent heating, the mercury evaporates and the gold precipitates as a thin coating. Caution is advised, as mercury vapors are extremely hazardous.

j. *Electrogilding:* Electroplating, galvanic gilding, which widely replaced the old technique of fire gilding. For the technique, see *Gilding.*

Giuliano, Carlo: Italian goldsmith and jeweler, died ca. 1912, famous for his special enameling techniques.

Glass: Glass is an amorphous solid material (not necessarily transparent) with a basic structure of a network of SiO4 tetrahedra connected by oxygen bridges as in silicates. Glass forms an undercooled molten mass and differs from the crystalline material in the condition of its fine structure. Thus glass has no definite melting point, but rather a softening or hardening interval. German: Glas.

Glass enamel: See *Email en resille sur verre.*

Glass-forming substances: Such substances include quartz and feldspar (which contains about 70% quartz), as well as boron oxide B2O3, phosphorus pentoxide P2O5, Aluminum oxide Al2O3, and lead oxide PbO. Quartz melts only at 1700 degrees Celsius. By adding other glass-forming substances with lower melting points, the melting point of the enamel is reduced to ca. 800 degrees Celsius. Fluoride is added as a flux. In addition to SiO2, other oxides (Sb2O3, As2O3, B2O3, P2O5) can function as glass-forming substances, since they also form such networks.

Glaze: Thin, glassy silicate overlays that provide the ceramic object with a smooth surface. In general, glazes, like enamels, are easily melted types of glass whose chemical contents vary within wide limits.

German: Glasur.

Glazes: With glazing colors that shine through (unlike covering colors), the previously painted-on colored layers can be made to shimmer through. This possibility of melting different transparent layers over each other so the lower ones shine through the upper ones also exists with enamel. German: Lasuren.

Glazing colors: They consist of metal oxides ground with about 80% flux. This carries the color pigments and bonds them to the surface. These colors are ground with turpentine or thick oil, painted on the smoothed surface and fired. The following burning intervals should be noted:

on porcelain	800 - 900 degrees C
on earthenware	780 - 800 degrees C
on glass	550 - 650 degrees C
on enamel	800 - 840 degrees C

Glitter enamel: A mixture of one part non-easily and two parts easily-melting enamel is melted on a surface. The hard-melting enamel remains in the easily-melting enamel when the latter melts, forming an uneven surface which is cut and polished. The particles of unmelted enamel act as enclosed bits of glitter and cause a glittering effect. German: Flimmer-Email.

Gloss firing: Another variation of further working after the first cutting. Transparent enamel in particular is suitable for gloss firing, as it would otherwise lose much of its glossy nature. The cleaned object is fired at about fifty degrees higher than normal for as short a time as possible, briefly and at higher temperatures because the enamel is just to be melted smooth, and with a short firing time the metals can scarcely oxidize. Any resulting *tinder* can be rubbed off with a solution of vinegar and salt after cooling. The brownish color tone of the somewhat oxidized wires can harmonize well with the whole play of colors and is thus left unchanged. German: Glanzbrennen.

Glycerine work: A decor is painted on a previously enameled surface with thinned glycerine. Then enamel that adheres only to glycerine is sieved on. After knocking off the excess enamel powder, the object is fired. German: Glyzerinarbeit.

Gold: A lustrous yellow metal, chemical

symbol *Au* from the Latin *aurum,* is the most flexible of all metals and can be beaten to gold leaf of 1/10000 mm thickness. It is also an unusually resistant metal, soluble only in chlorine solutions and aqua regia (1 part H_2SO_4, 3 parts HCl). Since it is so soft, it is usually alloyed with other metals. Pure (24-karat) and 18-karat gold can be enameled equally well.

Gold enamel plastic: *Email en ronde bosse:* This technique was practiced at the Burgundian court in the 15th Century and later, in Rudolf's reign (ca. 1600), in Bohemia. Cast figures are completely or partly covered with opaque or transparent enamels, producing colorfully enameled plastic forms.

Gold pit enameling: The pits in sheet gold are filled with opaque or, better yet, translucent enamels which, after firing, contrast very strongly with the non-enameled gold surfaces. German: Goldgrubenschmelz.

Granulation: (From Latin *granum*, meaning small grain). To a surface of the same metal (gold on gold, silver on silver), tiny gold or silver granules or balls are soldered. This often forms very effective ornamentation and heightens the effect of light and shadow on a decorative object. This technique was already known to the Etruscans and popular in the Middle Ages and the Baroque era. This technique has also been used in modern enamelwork.

Gray painting: See *Grisaille.*

Grisaille: Gray-on-gray painting; pictures painted in varying shades of gray. Particular depth effects can be achieved with it.

Groove enameling: A type of *pit enameling.* Instead of pits, thin grooves are cut into the metal and enameled. This technique is used most often on bronze and was previously used by the Celts. German: Furchenschmelz.

Guilloche: At the end of the 18th Century machines were being used for engraving. The entire metal surface is covered with regular lines, grooves, waves or bowed lines, called guillocheing. These are then partially covered with transparent enamel. Unfortunately, guillocheing has almost completely replaced artistic hand engraving.

Gum arabic: From the rind of certain types of acacias, a pectinlike water-soluble substance is obtained. Since it hardens, it is often used as an adhesive or coating.

German: Gummiarabikum.

Gum tragacanth: A plant gum obtained from various types of astragalus vetches. Mixed with water, it forms a dark, slippery slime which is used as an adhesive and covering.

Hairline cracks: Fine cracks in enamel layers, which come into being particularly when the enamel contracts more quickly than the metal surface after firing. German: Haarrisse.

Hardness: The hardness of a material is classified by which minerals do or do not scratch it, and which the material itself scratches. Friedrich Mohs developed a comparison table known as the Mohs table. Quartz, for example, has a hardness of 7, enamel and technical glass between 6 and 7, leaded glass between 5 and 6. German: Härte.

Hellot, Jean: French porcelain painter and color chemist, 1685-1766, who developed many underglaze and enamel colors.

Heuglin, Johann Erhardt: Silversmith and ornament designer, ca. 1672-1757, who lived in Augsburg and produced excellent traveling sets of silver as well as boxes, mirrors etc., decorated with inset enamel plaques.

High enamel: Thickly applied enamel decor. By hand or pressure, the contours of the pattern are applied with grease colors. The pulverized enamel colors are thickened with water and setting media (clay, colloids, silicic acid). With a brush or a painting horn into which the enamel color mixture can be pulled through via a cannula, the color is applied as in Engobe work on ceramics. The water-repellent grease colors prevent the enamels from running out or together. Blossoms or ornaments up to several millimeters thick can be modeled on glass by repeated application and firing. German: Hochemail.

History: There were already enamel artisans in Cyprus and Crete circa 1400 B.C. Sheet gold has been found there decorated with white and blue enamel. In Egypt too, *cell glazing* was already known. The enameling process reached Rome via Greece and spread to the Roman provinces. It was first introduced into Germany about the time of the Flavian Emperors (69-96 A.D.). In the Third Century the Celts had already mastered a type of *Champlevè technique.*

Pits in bronze forms were filled with enamel and fired. In this way swords, shields and equipment for horses were decorated, as well as jewelry. The enameling art reached a particularly high point in old *Byzantium* (500-800 A.D.). Here *cell* and *pit enameling* in particular were mastered. From Byzantium it became more and more significant to the Occident since the Sixth Century. In Germany the *Rhine and Maas School* (1000-1200 A.D.) dominated, in France the *Limoges* school. *Pit* and *cell enameling* were used since Roman times. In the French-Burgundian area costly pendants were made by *gold enamel plastic* work (chased or cast metal objects covered with enamel). In Limoges, where many religious articles were made by pit enameling, no pictorial plates were made since the end of the 15th Century, as it had been discovered that it was not necessary to divide and deepen the metal for the application of enamel, as it could be divided and applied to the entire metal surface. To this day this wet application is known as Limoges technique. The enameling art thus became a type of miniature painting. At the beginning of the 14th Century, under the Yuan Dynasty, the *Cloisonné technique* reached China from the Occident and Persia and developed to its highest point in the Ming Dynasty (1368-1644). The Chinese used it particularly to decorate hollow vessels such as vases, figures and religious articles, some of which attained considerable size. They also developed the *plique-à-jour technique* to previously unattained levels.

Höroldt, Christian Friedrich: German porcelain and enamel painter, 1700-1779, who worked at the *Fromery* enamel workshop in Berlin and later became a chief painter at Meissen.

Huaud, Pierre: He and his three sons were miniature and enamel painters in Geneva circa 1640, winning fame for their extraordinary works.

Hunger, Christoph Conrad: He worked in Meissen from about 1710 to 1750 as an enamel painter and gilder and helped to build up the industry in Vienna.

Inlay work: Already known in Egypt circa 2000 B.C., this is a preliminary step in enameling. Colored glass powder is thickened to a paste with adhesives and pressed into metal cells, where it hardens. This is not enamel, since the glass powder is not fired. German: Einlegearbeit.

Iron and steel: These enameled metals are used chiefly for kitchen utensils, cooking pots and industrial uses. Enameled iron has been used for household objects since about 1850.

Iron enameling: As early as 1764, cast iron utensils were coated with a glaze at the iron foundry in Königsbronn, Germany. In Sweden the enameling of sheet steel was first done in 1782, and in 1785 cast iron was enameled at the Lauchhammer works. In the next decades, various enameling firms worked independently, using their own empirical recipes, which the master enamelers kept in their possession. Conventional enameling is done in two layers. The *base enamel* with its bonding oxides (cobalt and nickel oxides) is melted first to form a good bond between glass and metal. Then the *covering enamel* is applied to the undercoated object in a second firing, giving the object its color and desired physical-chemical characteristics. Today it is also possible to enamel the objects in one firing. Here not only the economic aspects but also the thickness of the layers are important factors, for the thinner the enamel layer, the more resistant it is to being struck or dented.

Jünger, Christoph: Proprietor since 1767 of a workshop in Vienna where he manufactured enamelware. After his death in 1777, his brother Johann took over the firm and operated it until 1780.

Kothgasser, Karl Anton: Viennese porcelain and glass painter, 1769-1851, whose painted glasses rank among the most beautiful artistic creations of the Biedermeier era. He decorated great numbers of drinking glasses with landscapes, pictures of cities, portraits, flowers and inscriptions. The typical Kothgasser glasses, generally in curved conical form with rounded bases, were decorated with translucent enamel colors. He was one of the Biedermeier era's most famous enamel painters.

Kreussen: This Upper Franconian town, not far from Bayreuth, was an important center of stoneware production in the 16th and 17th Centuries. Articles decorated with relief applications and painted with colorful opaque enamels are especially striking. This was the earliest European use of enamel colors in ceramic decoration and was probably adopted from Bohemian glass painting. The most famous products are probably the apostle mugs.

Krüger, Jean Guillaume George: English enamel painter, born in 1728, who worked in Berlin and Paris.

Ku Yuch Hsuan: A Chinese porcelain decorated with enamel colors, its style adapted from Chinese Cloisonné work. It was first produced in the reign of Emperor *K'ang Hsi* and reached its peak between 1720 and 1760.

Lachenal, Edmond: This French ceramicist, 1855-1930, was the inventor of *Email veloute,* in which the surface of the glaze is etched in an acid bath.

Lalique, René Jules: This French goldsmith and jeweler, 1860-1945, was one of the most versatile craftsmen of his time. He was outstanding as an enameler and as a glass and jewel cutter. For his jewelry he used semiprecious stones and pearls as well as enamel.

Layer thickness: The thinner the enamel layer, the more resistant it is to striking and denting. German: Schichtdicke.

Lead content: Lead is often added to painter's enamel, since it increases the gloss and particularly the vividness of the colors. It is poisonous, though. German: Bleigehalt.

Limoges: Capital of the French Department of Haute-Vienne. The pit-enameled works made there from the 12th to the 14th Centuries are world-famous. In the 15th Century *painter's enamel* was developed there. A few artist families zealously guarded the secret of its production and prepared painter's enamel into the 17th Century.

Limoges Technique: In Limoges at the end of the 15th Century it was discovered that divisions and deepenings of the metal were not necessary for enamel application, as it could be applied over the entire metal surface. To this day this wet application is known as the Limoges technique (painted enamel). The enamel art became a kind of miniature painting. See *wet charging.*

Limousin enamel: Enamelwork from the Limousin workshop in Limoges. "Limoges enamel" and "limousin enamel" are often used incorrectly as synonyms.

Limousin (Limosin), Leonard: French enamel painter, 1505-1575, a member of one of the most famous families of enamel artists in *Limoges.* He was the king's court enameler.

Line technique: Decorative enamels tend to run into each other during firing. The resulting contours often damage the patterns. To prevent this, the outlines of the pattern are drawn on a pre-enameled surface with oiled crayons or waxed chalk. These lines must be unbroken so the enamels cannot mix. The resulting fields are then filled carefully with enamel powder, and the greasy lines repel the watery enamel paste. The oils burn off during firing and the enamels remain precisely divided and defined. To improve the contour contrast, the lines can be redrawn with dark painter's enamel powder and fired separately. German: Linientechnik.

Löwenfink, Adam Friedrich von: German porcelain and Fayence painter, 1714-1754, noted for his preference for gold and glowing enamel colors.

Maas School: Since the middle of the 12th Century, enamelwork from the Maas (Meuse) area, centered in Liege, have become more and more significant. Especially noteworthy is the goldsmith and enamel artist *Godefroid de Claire* of Huy on the Maas.

Mailly, Charles Jaques: This French enamel painter, 1740-1817, painted mostly still lifes and allegorical scenes in *Grisaille* technique, which were set into gold boxes.

Manufacturing: There are countless varieties of enamel. The processes by which they are made are complex and require much experience to produce them in balanced, harmonious colors and with unchanging melting properties. This has been done in glass factories for centuries. The basic substance of the enamel is ground *quartz* and *feldspar.* To these basic substances, fluxes are added, which considerably reduce their high melting points. *Adhesives* (binding oxides) and *darkening agents* are also added, as the melting flow would otherwise be transparent. The colors come from *metal oxides.* The proportions of the ingredients is decisive for the desired properties of the enamel. The precisely blended mixture is then melted at about 1200 degrees Celsius and then chilled in water. The resulting granules, the frit, are then ground in a mill with clay, pigments, and various agents, then mixed with water to a paste, forming a usable glaze sometimes called Schlicker or Schlempe in German.

Marbling: See *stirred enamel.*

Mary Gregory Glass: English name for inexpensive colorless or tinted glass painted with figures, usually children, in opaque white enamel. It was made in great quantities in Bohemia in the 19th Century and copied by enameler Mary Gregory of the Boston & Sandwich Glass Company in the USA, hence the name.

Masse, Jean-Baptiste: This French painter, engraver, and miniaturist, 1687-1767, painted enamel portraits of Louis XIV which were set into boxes decorated with jewels.

Matte finish: To achieve gloss/matte effects, all the portions of a completely enameled plate that are to remain glossy are covered with shellac. After drying, the piece is bathed in a weak solution of liquid acid. After the shellac is removed, interesting gloss-matte contrasts appear.

Matte-finish enamel: There is matte-melting enamel powder that melts at 800 to 820 degrees Celsius. If this temperature is exceeded, the surface becomes glossy. High-gloss surfaces can be treated with matte salts and be fired again at lower temperatures (800-820 degrees Celsius).

Mayer, Franz Ferdinand: Active as an enamel painter in England in the latter half of the 18th Century.

Melting point: Glass, enamel, and glazes differ from crystalline materials in their fine structure. They are amorphous. Thus they have no definite melting point, but rather a range of softening or hardening. Opaque enamel melts at about 820 degrees Celsius, transparent enamel at only 840 degrees Celsius. The high firing temperature results in good transparency. These melting properties are achieved by adding lead to normal glass at a considerably higher melting point. It is also possible to change the melting point of an enamel. For enamel that has a higher melting point, one can add a light-melting colorless fondant, and the melting point lowers. Enamel that has a low melting point is mixed in water with a few drops of hydrochloric acid added; it "hardens" and the melting point rises. German: Schmelzintervall.

Melting properties: These properties of enamels vary in both the temperatures and the firing time that are needed to melt them. Thus it is necessary to consider precisely in what order different-colored enamels are fired to get the desired results. In addition, certain enamel colors change or completely lose their color in a second or third firing at a higher temperature. Also, opaque colors can become transparent, and transparent colors opaque, at too-high temperatures. These circumstances must be carefully considered. German: Schmelzverhalten.

Metal oxides: See *pigments.*

Metal preparation: The surface that carries the enamel must be prepared specially, according to the kind of metal, by annealing, pickling, and cleaning, in order to decrease the chance of mistakes.

Metal thickness: A good-quality raw metal article is about 0.8 to 1 mm thick. For large plates and bowls, greater thicknesses are needed. If the metal is too thin, the article may bend and the enamel will break off. If a pattern is to be etched on, the metal must have an appropriate thickness. In determining the thickness, we must also mention *pit enameling.* Since enamel is placed in cuts in this process, naturally the thickness of the material is added to by that of the planned enameling, becoming 1 to 1.5 mm.

Meyer, Jeremiah: English miniaturist, 1735-1789, of German ancestry. He was miniaturist for Queen Charlotte and enamel painter for King George III.

Mica: The mineral forms elastic, flexible, platelike crystals that can be split apart. In the production of *window enamel,* such thin mica plates are placed under the open places in the metal. They are easily removed from the enamel after firing. German: Glimmer.

Mignot, Daniel: A French copper engraver and jewelry designer who worked in the late 16th and early 17th Centuries, especially in Paris and Augsburg. The backs of boxes, clocks, and ornaments decorated with enamel painting are typical of his work between 1596 and 1616.

Milk glass: White glass that appears milky because bone ash has been mixed into it, thus also called bone glass. It was especially popular in the 18th Century and was often painted with enamel colors. Today the milky effect is achieved with *darkening agents* such as titanium oxide. German: Milchglas, Beinglas.

Millefiori: Italian for "thousand flowers," a mosaic glass in which different-colored glass rods are bundled to make a floral pattern, melted together, coated, drawn into thin staffs and bundled and melted again. The finished glass staffs are then drawn and become thinner and thinner, yet they retain the exact color combination, so that when they are cut into plates, large cross-sections have the same pattern as small ones. These plates, set side by side on enamel or glass and melted in, or enclosed in paperweights, provide a colorful mosaic pattern.

Millefiori inlay work: Millefiori plates are glued cold on a metal surface or melted into *fondant.*

Miniature portrait painting: See *enamel painting.* They are painted in metal oxide colors, usually on a white enamel background, and fired. This technique was especially popular in the Rococo (1730-1780), Classic (1770-1830) and Biedermeier (1815-1848) eras.

Mixing: Mixing an enamel depends on the chemical character of the metal to be enameled and the intended use of it. The basic type of enamel has the following composition: 33.5% silicic acid, 41.5% lead oxide, 17.4% potash, 7.6% boric acid. German: Zusammensetzung.

Mohn, Samuel: 1762-1815, and his son Gottlob Samuel Mohn, 1789-1825, were Saxon porcelain and enamel painters who devoted themselves to painting cylindrical glasses with views of cities in translucent enamel painting. The panorama glasses of these two hardworking artists, who produced them to order by a duplicating process, were desirable travel souvenirs of the spas visited by the upper classes at the time.

Mosaic: Coarsely ground enamel, also known as splinter enamel. Strewn over a pre-enameled surface, it remains raised after firing.

Moser: This German enameler and goldsmith emigrated to England. All the work on his boxes was done by himself. They stand out for their excellent enameling and a great range of glowing colors.

Muffle colors: See *Vitrifiable colors.*

Muffle oven: Electrically heated kilns, thermically controlled, are used to fire enamelwork. The nucleus is the muffle of fire-brick in which the articles are fired. German: Muffelofen.

Musive work: inlaid mosaic work. German: Musivarbeit.

Niello: Latin *nigellus* + blackish; this technique can be called a predecessor of real glass enamel. Niello is a black mixture of silver, copper, lead, sulfur and borax, which is rubbed into engraved or etched designs and fired at low temperatures. The recipes for this black mixture have varied in different cultural epochs, and even today goldsmiths keep their own tested recipes secret. The technique was already known in earlier times and was particularly popular in the 11th and 12th Centuries and the Renaissance. In the 19th Century richly decorated articles, known as Tulaware, came from the Russian city of Tula.

Nikolaus von Verdun: An enameler and goldsmith of outstanding importance. Between 1181 and 1205 he became a master of *pit enameling* in particular. His most important works are found in Klosterneuburg Monastery near Vienna (altar antependium) and in the Cologne Cathedral (great shrine of the Three Kings).

Ninsei: The artistic name of the Japanese painter and master potter Seiemon Nonomura, who worked in Kyoto between 1630 and 1695. His enamel paintings on glazed stoneware are most important.

Opalescent enamel: The name suggests the opal, the semiprecious stone that is half opaque, half transparent. This opal-like effect of letting the background shine through easily can be achieved only by a subsequent firing at about 400 degrees Celsius. German: Opales Email.

Opaque enamel: Nontransparent enamel, darkened with tin oxide or phosphate (see *darkening agents*) and completely covering the metal surface.

Orange peel: When the enamel is not fluid enough or the firing temperature in the oven is not sufficient to let the enamel flow properly, the surface after cooling resembles the peel of an orange. German: Orangenhaut.

Ottonian enamel: Only between 936 and 1002 did gold pit enameling in the form of *sink enameling* and *full enameling* gain in importance north of the Alps. The city of Trier played an important role. The works

show a light brightness of color that indicates their *Byzantine* influence.

Overfiring: At normal firing temperatures the enamel melts without overspreading the intended pattern. In overburning, temperatures about 100 degrees Celsius higher than normal are chosen to make the colors mix with each other. German: Uberbrennen.

Overglazing colors: see *vitrifiable colors, enamel colors.*

Paillon: A technique known since the 15th Century. Various shapes (flowers, animals, figures) are cut out of gold, silver, or platinum foil, glued to an enameled surface with *gum tragacanth* and fired. They are then covered with fondant or transparent enamel and fired again. After cooling, the article is smoothed and polished. In the 18th Century, J. Coteau developed the technique further in France by melting the metal foil between two translucent layers of enamel.

Paint oil: Binding media for enamel colors are sandalwood oil, oil of cloves, or the finest turpentine.

Painter's enamel: Limoges enamel. This method was first used at the Burgundian court around 1400 and then developed in Venice and particularly in Limoges in the 15th to 17th Centuries. A surface is first covered with an opaque solid-color enamel layer. This forms the background for the painting. With very fine enamel powder (600 mesh) mixed with water, the picture is painted as with paint. Only after all the water has evaporated from the article can it be fired at 850 degrees Celsius. For a multi-colored enamel picture, several layers must be applied and fired in turn, since every enamel color has its specific firing temperature. It was important to also apply the colors in the right order, since the colors that melted at the lowest temperatures could only be fired last. Beginning with the highest temperatures, the colors were fired in order, each color at a lower temperature than the last. The oven could not be too hot, or the enamel would flow too much and blue the design. Parts of the metal often remain free in this technique. On the other hand, in enamel painting metal oxide colors are applied to a white enamel background (miniature portraits, clock decorations, as in the 18th Century). German: Maleremail; French: Email des peinters.

Pardoe, Thomas: English ceramic painter, 1770-1823, who worked as an independent

enamel painter in Bristol after 1809 and was well-known as a flower painter.

Patina: A layer that forms on the surface of metals through oxidation, is also produceable artificially, and is sometimes used as a decorative element.

Pea green: Green enamel color, introduced in Sèvres in 1756 and copied in Chelsea and Worcester in 1768. German: Erbsgrün.

Peeling: Metals and enamels expand and contract differently when heated or cooled. For that reason the expansion coefficients of the (metal) enameled surface and the enamel must be attuned to each other, otherwise enamel can peel off. Slight differences can be overcome by a certain elasticity of the enamel. German: Abspringen.

Pénicaud: French enameling family of the 15th to 17th Centuries, in Limoges; the most famous members were Nardon, circa 1470-1543, his brother Jean I, active between 1510 and 1540, Jean II, before 1588, and Jean III, in the latter half of the 16th Century.

Petitot, Jean P. the Elder: French-Swiss miniature painter, 1607-1691, from Geneva. Among other cities, he worked in London and Paris, became court painter to Louis XIV, and then returned to Switzerland. He created miniature portrait masterpieces in enamel painting. His son Jean-Louis (1653-1699) carried on his father's business.

Photo enamel: In the early days of photography, glass plates were covered with gelatine for photographing, with light-sensitive grains of silver halogen embedded in it. By exposure to light and subsequent developing, the picture was fixed in the form of silver crystals. The gelatine layer was then loosened from the glass and transferred to a pre-enameled surface. After drying, it was fired; the organic gelatine layer burned away and the silver crystals sank into the enamel layer. After this firing the "photo" was softly colored and partially retouched. German: Fotoemail.

Pickling: Inorganic impurities are removed from a metal surface with the help of acid mixtures. First the surface must be degreased, since oils and fats prevent the effect of pickling acids. German: Beizen.

Pigments: All substances that color the enamel flow are regarded as pigments. They are generally metallic compounds such as oxides, spinels, conventional pigments, and heavy metal ions. Countless color variations are possible. A few examples:

Cadmium sulfide	Yellow
Cobalt oxide	Blue
Chromium oxide	Green
Selenium compounds	Red
Iron (+ 3) oxide	Blue-green
Gold chloride	Ruby red

Pinchbeck: See *Tombac.*

Pit enameling: In this technique, holes are made in a metal plate by cutting, piercing, chiseling, sawing, or soldering a cut plate to a baseplate. Enemal is placed in these holes and melted. After cooling, the piece is smoothed and gilded. The enameled surfaces of the pits stand out from the background through their glowing colors, unlike etched pits. Experienced goldsmiths and chasers cut the pits out of thick material with pointed or small flat chisels or punch them into the surface of much thinner sheet metal. The etching process is somewhat more laborious and difficult. This technique of pit enameling was already known by the Celts in ancient Roman times (as a special form of furrow enameling) and reached its peak on the Maas and lower Rhine and in Limoges in the 12th and 13th Centuries. German: Grubenschmelz. See also *Champlevè.*

Plastic enameling: A silver, copper, or gold figure is partly or wholly covered with opaque or transparent enamel layers, resulting in colorful three-dimensional enameling. German: Emailplastik; French: Email en ronde bosse.

Plating: Process of surface covering. Layers of a precious metal are applied to a base metal mechanically, by cold-rolling or welding, or by galvanizing. German: Plattieren.

Polishing: If an enamel layer is to be brought to a high polish without the usual gloss firing, it is done by polishing with bits of lead, cork, lindenwood or pumice. Finally, the surface is buffed with a felt disc. German: Polieren.

Pontypool Ware: Lacquered and painted metal goods made by the Allgood family of Pontypool, England. After his father, John Allgood, had tried in vain to apply lacquer painting to sheet iron, Edward Allgood solved the problem around 1730. He tinned the uneven iron surface, making it smooth and receptive, before applying a lacquer particularly insensitive to heat. Soon he founded the "Pontypool Japan Works," whose products were very popular because of their durability. This success led to the founding of other firms such as J. Bartlett and Sons in Bristol and John Baskerville in Birmingham. Allgood's products were superior to all competition in Europe other than those of the Stobwasser firm in Germany. They were often sold as substitutes for enamel, for the mass-produced goods from Birmingham, Bilston, and London were less expensive. These hand-painted pieces, of course, had their price. The quality of Pontypool Ware sank at the end of the 18th Century and the firm was closed down in 1822.

Porcelain: The name presumably comes from Marco Polo. The material is essentially kaolin, a particularly fine white clay, the name of which presumably comes from its discovery site in China, Kao-Ling. There are two types: soft and hard porcelain (1300-1500 degrees Celsius). Soft porcelain is produced mainly in China, Japan, and England and has only a small kaolin content. Hard porcelain is produced chiefly on the Continent of Europe, particularly in Germany. The high kaolin content (50-60%) and the feldspar glaze give it its extraordinary surface hardness and density. Another type of soft porcelain, known as bone china, is so called because of its large bone-ash content (50-60%). The kaoiln content is considerably less than in hard porcelain. Porcelain that is fired unglazed is called biscuit porcelain. The fired piece is nevertheless watertight, though it has a dull matte surface. German: Porzellan.

Porcelain painting: The color decor is applied either under or over the glaze. In underglaze painting, high-firing colors are used—metal oxides that can stand high temperatures, such as cobalt or iron red. These are applied directly to the material, covered with glaze and then fired. The oldest over- or underglaze technique was painting with enamel colors whose range was rather limited. These colors generally stand out physically from the surface glaze, since they cannot be melted at the high temperatures needed to melt the glaze; thus they allow a bonding of color and glaze. But on soft porcelain they can melt into the glaze. Overglaze colors (also called muffle colors or melting colors) are lead or borax glazes colored with metal oxides. The flux contained in these colors makes them melt with the background glaze at low temperatures (600-800 degrees Celsius) to give a flat surface after glazing.

Pratt, Felix: English potter, 1780-1859, from Fenton. He made figures on stoneware vessels with relief decor, as well as on terracotta since the early 1840's, painting it with glowing enamel colors.

Precious metals: Gold and silver cause strong color changes in enamel, quite unlike copper. Thus transparent enamel has the best effect on these precious metals, as the reflective quality of the basic material is most effective. To imitate the color effect of gold and silver, gold or silver foil can also be placed on the article and then covered with transparent enamel. German: Edelmetalle.

Preissler, Daniel: 1636-1733, and his son Ignaz Preissler, 1676-1741, were painters of German birth who worked mainly in Breslau and Vienna, painting Bohemian glass as well as porcelain (Meissen, Viennese) with enamel colors and black solder.

Preparation of enamel: If one buys lumps instead of powdered enamel, they must first be pulverized in the mortar and ground finer. The enamel is then divided for specific purposes with sieves of definite granule sizes and kept in glass bottles. Before application, the enamels must be washed or liquefied until the water runs clear. This is especially important for transparent enamels; only with opaque enamels may the water be clouded. The strictest cleanliness is required in working, as dust disturbs the whole effect of enamelwork. It is also sensible to fire the enamel in several layers, since many types of enamel become clouded or milky when they are applied too thickly.

Pressing: Pressing sheet metal between two forms, this stamping from stone, bronze, or iron has been known since the Second Century B.C. German: Stanzen.

Properties: Enameled objects unite the advantages of glass with the strength of metal. The thinner the enamel layer, the less fragile the glass coating, the more elastic the enamel. It protects the surface and is especially resistant to cooking, acids and other chemicals, fading, changing temperatures and weather, protects against corrosion and much more. The networks of silicic oxide tetrahedra contain alkaline substances that determine the physical and chemical properties of the enamel. A series of oxides also are present in glass that have no glass-forming qualities themselves, but the chemical and physical properties of the glass are considerably dependent on them. According to the intended use, the desired properties of the enamel also can be influenced by mixing the ingredients. Lead oxide (PbO) increases the specific gravity, refracting qualities and polishing potential of the glass (leaded glass). Easily melting glass contains considerable amounts of alkaline oxides, especially Na_2O. As the amount of SiO_2 and B_2O_3 increases, the glass or enamel becomes harder to melt and thus more resistant to temperature changes and chemicals. German: Eigenschaften.

Punching: The punched pattern is struck into the metal negatively. The process is used particularly to apply small ornaments, frames, and details and to fill out a single surface. German: Punzieren.

Quartz: Silicon dioxide, SiO_2, a mineral that is colorless or colored by various metallic ions, often used as a decorative stone (rock crystal, amethyst, rose quartz, agate, etc.). It is the world's second most common and widespread mineral, after the feldspars, and one of the main ingredients of porcelain, glass, glazes, and enamel. SiO_4 tetrahedra are linked by oxygen bridges and form spatial structures, thus functioning in glass formation. German: Quarz.

Randall, Thomas Martin: This English painter and porcelain manufacturer, 1786-1859, founded the Robins & Randall enamel-painting works in Spa Fields, near London—mainly decorating white porcelain.

Ravenet, Simon-Francois: A French designer and engraver, 1706-1774, later the chief designer at Battersea who created the *Battersea enamel* used primarily for royal portraits. religious and mythological, as well as generic scenes.

Raw enamel: Enamel frits sold by their manufacturers in large or small lumps. German: Rohemail.

Raw materials: The most important raw materials for glass and enamel production are the minerals quartz, SiO_2 (in the form of quartz sand, rock crystal or quartzite), $CaCO_3$ (chalk or dolomite), and potassium carbonate, plus borax minerals (borax, boracite, sassoline). German: Rohstoffe.

Raw surface: Raw objects, plates, etc., the pieces to be enameled. German: Schmelzträger.

Relief: A design is made three-dimensionally on a surface. According to the projection of the design, it can be classified as low or bas-relief, high or haut-relief, or half-relief.

Relief enamel: This technique is an offshoot of *Cloisonné*. As there, flat fillets are bent to form patterns and soldered to the surface. But the cells are only partially filled with enamel powder, the rest remaining empty. After repeatedly filling and firing certain cells, the excess enamel is ground off and polished and the metal is then gilded galvanically. The enameled motifs form a relief that stands out from the gold background.

Relief enameling: Also called *deep-cut enameling* or *email de basse taille*. Since strongly reflective silver was preferred to copper as a background, this type of enameling is generally called *silver relief enamel*.

Reymond, Pierre: French enamel painter, who died circa 1548, a member of one of the renowned enameling families of *Limoges*. He usually painted *Grisaille*.

Rhenish pit enameling: In the 12th Century, the Maas School and the Lower Rhine School, centered in Cologne, were of great artistic importance in the development of the pit enameling technique. At first this art was practiced chiefly in the great cloisters in the Diocese of Cologne, but soon city workshops came to the fore, and many of the famous artists of the time were laymen. Among the important names are Reiner von Huy; *Godefroid de Claire,* whose works include the Alexander Reliquary in Brussels, the Heribert Shrine in Deutz, and the Cross of St. Omer; and *Nikolaus von Verdun,* who made the enameled Klosterneuburg Altar in 1181. Hugo von Oignies worked in Namur Abbey and Eilbert in Cologne; the latter is the creator of the portable Guelph Altar.

Roman enamel: Enamel was scarcely produced in Rome itself during the first half of the Empire. The significant pit-enamelwork was done in the Roman provinces, particularly for the use of the army.

Royal blue: See *Derby blue.*

Running enameling: Steep-sided vessels have granules of colored enamel placed along their upper rim on a hard-melting enamel surface. During firing, gravity makes the colored enamel run down the sides and gather to form a colored rim, leaving tracks on the way down. German: Laufschmelz.

Salt, Ralph: English potter, 1782-1846, from Hanley, who decorated his figures with enamel and luster colors.

Samson: The Samson factory in France copies boxes from *Battersea* and *Bilston* as of 1860. The first products were marked with the letter S, but this was soon eliminated.

San ts'ai Ware: Chinese for "three colors"; a new technique of three-color painting (yellow, purple, and green) that appeared in the Ming era under Emperor Chia Ching (1521-1566). The enamel colors were applied to stoneware or porcelain pieces in fields limited by raised lines of clay. A derivative of the Chinese *Cloisonné.*

Schaber, Johann: A German painter, 1621-1670, who worked in Nu'rnberg, making delightful little drinking glasses, usually with winged bases. They were painted in transparent enamel colors, fine nuances of black solder, and gold painting. He also decorated Fayence articles.

Scharf, Johann Gottlief: This Russian box maker, who died in 1808, made round gold boxes decorated with enamel and precious stones for Catherine the Great in St. Petersburg.

Schindler: A Viennese enameler of the latter half of the 18th Century, he supplied the Congress of Vienna with enameled boxes of outstanding quality.

Scratching technique: See *Sgraffito.*

Seuter, Bartholomäus: A German painter, 1678-1754, and member of a well-known Augsburg engraving family. He decorated Fayence pitchers and Meissen porcelain with enamel painting and gilding.

Sgraffito: Scratching technique: An enameled object is completely covered with enamel powder in a contrasting color, parts of which are then removed with a brush. After firing, a bright-colored pattern emerges.

Shorthose, John: An English potter, 1768-1828, from Hanley, who produced cream-colored stoneware decorated with enamel painting for export.

Sieving: A method of applying enamel to a surface. The enamel powder is dusted through a sieve onto a surface sprayed with gum tragacanth solution. It is the safest and simplest way to apply an even coat of enamel to a large surface. German: Aufsieben.

Silk-screen printing: Grease colors are picked up on silk paper from lithographic stone, laid on the article to be enameled like an adhesive picture with the colored side down, and the design is rubbed on. The silk screen can be removed; the grease color remains on the enameling surface. Then fine colored enamel powder is dusted over it and sticks to the grease color. After drying, it is fired. In this process a separate firing is needed for each color; see also *glycerine work.*

Silver: A gleaming white metal (chemical symbol *Ag* from the Latin *argentum*), harder than gold and second to it in flexibility. Hydrogen sulfide, phosphorus and hydrogen sulfide in the atmosphere cause a dark coloration of the surface. Silver is also alloyed. Pure silver (100% pure) is especially suitable for enameling, as it does not oxidize and thus forms no tinder, and can be heated almost to its melting point (961 degrees Celcius). Sterling silver is an alloy and melts at 893 degrees Celcius. Thus no enamel can be used that requires a higher melting temperature; also, many enamel types flake off or lose their brightness. German: Silber.

Silver foil: Very thin leaves of pure silver that can be fired between two layers of transparent enamel.

Silver relief: The decor is cut into sheet silver as a low relief and then covered with translucent silver enamel. The light is affected differently by cuts of different depths; the deeper the cut, the darker the enamel appears.

Sink enameling: A term applied to *cell enameling* when the *Cloisonné* is sunk into the metal plate. A pit is made in a thin metal plate by flat punching; bridges are set in this pit to form the design and then filled with enamel and fired. The cell-enamelwork is then on the same plane as the non-enameled part of the metal, which usually forms an outline around it. This is a combination of cell and pit enameling, sometimes called email mixte in French. German: Senkschmelz.

Smalte: A potassium cobalt silicate, a flux-rich clay mixture used as an underglaze color as well as for blue glass coloring. German: Schmalte.

South Staffordshire Enamel: Made mainly in the centers of Bilston and Wednesbury, its large production included boxes, tea chests, trays, cases, small bottles, and candlesticks. Most pieces were made by the *transfer process.* In the beginning, the transfers were painted over with transparent colors that let the design drawings show through. Later they were painted in and covered with opaque enamel colors. The use of colored fonds as of 1760 is typical of South Staffordshire, as is the surrounding of the pattern with relieflike enameled tendrils and pearls.

Spatula: A simple tool for applying enamel. German: Betragstift.

Stamping: A technique related to pressing, in which a pattern with all its details is stamped into one side of a metal piece. German: Prägen.

Steel pressing: In this process patterns are engraved in steel plates. The enamel colors are pressed into the cuts as a thick paste and the excess is scraped off. The color remaining in the cuts is removed with paper and rubbed off onto the article to be enameled. Copper plates were used formerly, but they wore out sooner and did not allow as many transfers. The steel-pressing process allows the transfer of several colors without firing in between, for after one color dries, the next one can be applied and all colors can be melted in a single firing. German: Stahldruck.

Steel pressing transfer: The process is done as in steel pressing, but the color is transferred to the article not on paper but by means of a gelatine foil.

Stencil painting: For series production and also single items, all the parts of a pattern are cut out of water-resistant cardboard or oiled paper so that the openings in the stencil are linked by small bridges. The finished stencil is placed on the pre-enameled piece. Using a well-moistened brush, one paints dots from the edges of the cuts to the middle of the stencil openings; the stencil must not be moved. After the color has dried, it is fired. German: Schablonen-malerei.

Stencil technique: A surface is enameled in one color. Then a paper stencil is cut, placed on the enameled surface, and dusted with a contrasting color. After the stencil has been lifted off carefully with forceps, the piece is fired.

Stiegel Glass: Green, blue, or colorless glass, first produced in America, that was blown into daisy-patterned forms and then engraved or painted with enamel colors.

Stölzel, Samuel: German firing master, potter, and color chemist from Meissen, who died in 1737.

Stoneware: Unlike *earthenware,* it is fired heavily. The whitish-gray material is impervious even without a glaze. German: Steinzeug.

Surrey Enamel: The name comes from the mistaken assumption that this technique originated in the English county of Surrey. It is a derivative of the *Champlevé technique used on enameled brass objects, very popular in Europe, and especially in England, in the 17th Century. The pits to hold the enamel were not cut in the material afterward but were cast in and thus present on the cast objects. Then the usual Champleve/* technique was used.*

Swirled enamel: Different enamel colors are applied to the surface and then put in the oven. Through the open door the molten enamel is stirred with a tool. The result depends very much on chance but produced interesting colored effects. German: Quirl-Email.

Technical application of enamel: Along with wet application by dipping in liquid or semiliquid mixtures or by spraying, dry application is done by sieving, and more recently by layering in an electrostatic field with wet or powdered enamel. In the process, 95 to 98% of the applied enamel falls on the object to be covered.

Test samples: Since enamel colors vary not only in firing time and temperature, number of firings and reaction to surfaces, but also in manufacture, it is vital to make test plaques. The plaque is divided into zones and enamel is applied directly to the metal on a surface of fondant, on metal foils, on hard and soft enamel, and various applicable techniques are tested. German: Probeplatten.

Theophilus Presbyter: He presumably lived in the abbey of Helmershausen near Paderborn around 1100 and wrote the "Schedula diversarum artium," the most important technical art textbook of the Middle Ages. The first written description of enameling comes from Theophilus.

Thick oil: Thickened turpentine, lavender oil, oil of cloves, or other oil. German: Dicköl.

Tiffany, Louis Comfort: Painter and glass designer, 1848-1933, a renowned Art Nouveau artist. In addition to his famous glass articles, he also created jewelry decorated with jewels and enamel.

Tinder: An oxide layer that forms on metal as a black crust during firing. German: Zunder.

Tinder protection: To keep the front of an object clear of counter-enamel while firing, it is covered with a protective layer to prevent the copper from oxidizing. This protective coating can be pulled or brushed off after firing. But simple deacidifying and cleaning also suffices. German: Zunderschütz.

Tombac: A reddish to gold alloy of copper and zinc, also called pinchbeck. German: Tombak.

Toreutik: German term for the art of working metal, particularly embossing but also punching, chasing, etc.

Tou ts'ai: Chinese for "warring colors." Ceramic pieces with a blue underglaze are decorated after firing with enamel colors on the glaze. The underglaze painting serves as an outline, over which the enamel colors are applied in a thin translucent layer. This style was developed in China during the Ming era.

Tracing stencil: Since the 17th Century, this technique of ceramic mass production has been used, as it is by handicrafters today, to transfer designs. A box painted with patterns is cut along the design lines with a pattern-cutting wheel and thus equipped with fine holes. Then it is set on the pre-enameled plate and pulverized charcoal, graphite powder, or pumice is rubbed through the holes directly onto the surface. The resulting dotted lines are painted over with enamel and fired. German: Durchpausschablone.

Transfer decor: Decals, transfer pictures, decorative pictures; the color pigments of the pattern are put in an organic gelatine and transferred to the paper carrier. To transfer the pattern, the paper carrier is moistened with water for several minutes. Thus the pressed gelatine foil is loosened from the paper and can be applied smoothly to the object. As long as the gelatine is moist, it can be pushed over the object until the proper place is reached. Pressing removes air bubbles from under the gelatine. After drying, it is fired. In the process, the gelatine burns up and the enamel colors are attached to the surface.

Transfer printing: At first, one or more engraved and colored copper plates are used, as in copperplate engraving, to make a one- or multicolored pressing, on which a thin transfer medium, usually paper, is pressed. To bind and protect the colors, the whole design is covered with a thin layer of gelatine. Such design carries can also be bought and can be applied to the surface of the object after being dipped in water like a wet transfer. In the subsequent firing, the organic gelatine layer burns off and the color pigments melt into the glazing. German: Abziehbildverfahren, Umdruck-verfahren.

Translucent enamel: See *transparent enamel* and *Basse-taille*. The term is sometimes used in reference to *window enamel* as well.

Transparent Cloisonné wire enamel: A very laborious technique demanding much ability and patience. The Chinese are the masters of this technique. For small articles a thin copper piece, as in the Cloisonné technique, is marked, rolled flat, and wires bent to form patterns are soldered on—which must be attached solidly to each other as well. Then the cells are filled with transparent enamel and fired. This process is repeated until the cells are filled. After being smoothed down and glazed to a gloss, the copper object is dissolved in acid. For larger and particularly more complex objects, a plaster form is cast and the wires are soldered over it as a fine network of wires bent to shape. Now the plaster form is dissolved with sulfuric acid and the wire filigree framework is cleaned and silvered. In very intricate work, the individual cells are now filled with enamel powder. The various transparent colored enamels are held on with gum tragacanth and pushed through the fine openings in clumps with forceps. When the piece is fired, the enamel sinters together and leaves half-closed cells. The openings must be filled and fired again and again until a complete glass surface is formed. Naturally enamel will also run over the wires. The excess enamel is removed and the transparent enamel given its glowing quality in one last gloss firing. Usually the wires are then gilded galvanically. German: Transparentes stegemail.

Transparent enamel: As opposed to opalescent or opaque enamel, transparent enamel lets the metal surface shine through and add to the effect. If this surface has been polished to be mirrorlike or decorated, fiery background effects can be achieved. This effect is especially intensive with pure silver or gold.

Underglaze colors: These colors are painted onto the hot material before glazing and then fired at about 1400 degrees C, uniting with the surface and glaze. In the 18th Century only cobalt underglaze blue was known to withstand these temperatures. Later chromium green, brown, and black (based on iron and manganese oxides) were added. At the end of the 19th Century, modern color chemistry made it possible to add a series of yellow, brown, and red color tones to the "sharp-fire colors."
Blue - Cobalt
Purple or brown - Manganese
Green - Copper
Yellow - Antimony
Brick red - Iron
German: Unterglasurfarben, Scharffeuer-farben. At Fayence they are also called Inglasur colors.

Underglaze painting on enamel: The colors are painted on a pre-enameled surface but remain matte after firing. In a further work process the painting is covered with transparent enamel and fired after drying. Only then do the colors become glossy. German: Unterglasurmalerei auf Email.

Unglazing: See *aging*.

Use: Enamel serves as surface protection against atmospheric and chemical effects (industrial and technical uses), to attain smooth metal surfaces (i.e., cooking pots, kitchen utensils), usually of sheet or cast iron, and for decorative purposes (jewelry, ornamental objects), for which copper and tombac are usually used, along with gold and silver.

Vachette, Adrien-Joseph-Maximilian: A French jeweler and boxmaker who died in 1939. He decorated his boxes with plaques of porcelain, tortoiseshell, and enamel as well as enamel-color portraits.

Varnish: Solutions of resinous substances in oil (linseed oil, for example) or alcohol. German: Firnis.

Varnish firing: Brown varnish (incorrectly also called email brun) is used to partially cover a copper plaque which is then fired. The unpainted areas are gilded or silvered. The brownish varnished portions offer a nice contrast. This technique was used particularly in the 12th and 13th Centuries. German: Firnisbrand.

Vermeil: Fire-gilded silver.

Vestier, Antoine: A French painter, 1740-1824, who copied paintings in enamel.

Vitrifiable colors: Muffle, glazing, or overglazing colors for porcelain, earthenware, enamel, or glass. A glass containing lead or borax is melted and colored with metal oxides, painted on the surface of the fired article, and melted in the muffle oven. The firing temperature depends on the burning temperature of the glassy surface and is between 500 and 900 degrees Celcius for enamel and glass, about 800 and 1000 degrees Celcius for ceramics and porcelain. German: Schmelzfarben.

Washing: The enamel surface is covered with more or less large pores that fill up with bits of ground-off material, producing gray spots. It is urgently necessary to remove these bits with a glass brush under running water—to wash them out. A great deal of the effect of the finished enamelwork depends on this work process. Only now should gloss firing be done. German: Auswaschen.

Wednesbury: An English town in the southern part of Staffordshire. Here in 1776 Yardley began production of enamelware. Early works were decorated with shepherd or gallant scenes. Most inexpensive souvenir boxes probably were made in Wednesbury, where production continued until 1840.

Wet charging: See also *Limoges technique.* For complicated patterns where precision is required, wet application gives the best results. The colors can be applied side by side with spatula or brush without running together. German: Nassauftrag.

Window enamel: Pieces are cut out of a metal plate with a jeweler's saw. The framework can also be made up of metal bridges. Then the metal is backed with copper foil or laid on a mica plate so the enamel will not melt onto the firing underlay. Now the openings are filled with transparent enamel and fired. This process is repeated until the openings are filled and fired smooth. The protective underlay is removed from the finished piece, the back is smoothed and polished. The effect resembles that of stained-glass windows. German: Fensteremail.

Wolfers, Philippe: Belgian goldsmith and handicrafter, 1858-1929, who worked with glass, enamel, and precious metals. His refined and imaginative decorative works made him, along with Lalique, one of the most important revitalizers of the art in Europe.

Wu ts'ai Ware: Chinese for "five colors"; the term for porcelain painted in red, green, yellow, and black enamel colors on a blue underglaze. This technique appeared in the Ming era under Emperor Ch'eng Hua, 1464-1487.

Yardley, Samuel: In the latter half of the 18th Century he founded the enamel industry in Wednesbury, Staffordshire.

Zinke, Christian Friedrich: An enamel painter, 1684-1767, who was born in Germany and worked in England as of 1706, creating many small pictures in glowing enamel colors; they were often set in boxes.

TRANSLATOR'S NOTE
Like practically any other specialized field, enameling has more specific terminology in German than in English—or so it seems when one translates! When I have not been able to find an equivalent English term in literature on the subject, I have translated the German terms literally, and have added many German terms at the end of their descriptions in the glossary. I hope you will understand what is meant.